Bureaucracy, Bankers and Bastards: a farmer's story

Kathryn Spurling was born in Brisbane. Being a restless soul and from a military family, she enlisted in the Women's Royal Australian Naval Service (WRANS). Having married an RAN Electrical Engineering Officer, she and the family moved interstate and overseas. Having completed a Masters degree while living in the USA, she commenced a PhD in history at UNSW at the Australian Defence Force Academy, Canberra, in 1996. This was granted in 1999. Over the ensuing years, she tutored first year History and Strategic Studies and guest lectured in Australian, naval, and women's history. From 2004 to 2011 she was a Visiting Fellow at UNSW@ADFA as well as lecturing at Denmark's Defence College; University of Massachusetts; NATO Headquarters, Brussels; and Spain; and was a Summer Fellow appointed to the United States Military Academy (Westpoint), New York. At the beginning of 2012 until 2014, Kathryn was appointed to the Australian National University as a Visiting Fellow until she accepted an Adjunct Research Fellowship with Flinders University, which she continues to hold.

kathrynspurling.com

Also by Kathryn Spurling
Cruel Conflict: the triumph and tragedy of HMAS Perth
A Grave too Far Away: a tribute to Australians in RAF Bomber Command, Europe
Inspiring Australian Women
The Mystery of AE1: Australia's Missing Submarine and Crew
HMAS Canberra: Casualty of Circumstance
Abandoned and Sacrificed: the tragedy of the Montevideo Maru
Fire at Sea: HMAS Westralia 1998

Kathryn Spurling

Bureaucracy, Bankers and Bastards: a farmer's story

Acknowledgements

With thanks to William 'Bill' Mott for his patience, honesty and determination in 'not allowing the bastards to win'.

Thanks also to Stephen Matthews and Ginninderra Press for having the courage to publish *Bureaucracy, Bankers and Bastards: a farmer's story*.

Bureaucracy, Bankers and Bastards: a farmer's story
ISBN 978 1 76109 034 9
Copyright © text Kathryn Spurling 2020

First published 2020 by
GINNINDERRA PRESS
PO Box 3461 Port Adelaide 5015
www.ginninderrapress.com.au

Contents

Preface	7
Introduction	9
One	11
Two	18
Three	31
Four	41
Five	49
Six	56
Seven	66
Eight	73
Nine	80
Ten	90
Eleven	96
Twelve	105
Thirteen	115
Fourteen	125
Postscript	131

This book is dedicated to Australian agricultural and rural communities.
Your true value is too often lost on others.

Preface

Like most Australians, I had observed the erosion of national values and beliefs, and of ethical behaviour practised by big business and institutions. 'Greed is good' appeared to have taken precedence over basic interest in, and concern for, human welfare. Whatever corporations and governments could sell in the name of short-term profit was embraced. This myopic view had and has, increasing repercussions for farming communities, water and food security and the fragile Australian ecosystem.

It was by pure chance that I met William 'Bill' Mott, and his story alerted me to the reality. It was a sharp learning curve to physically and emotionally explore the world of Australian agriculture and farming communities. Sitting at the bar in an outback country pub listening and observing was an education. Travelling the broad horizons of a sun-baked Queensland, and walking the all-but-deserted streets of what were once thriving country towns, was fundamental to understanding the dilemma. Comprehending 'pinched wheat' and 'moisture profile', the advantages and disadvantages in sorghum, wheat and chickpea crops, as well as black and red soils, was challenging. I thought about this when seated in the air-conditioned, comfortable main hearing room of Parliament House as I witnessed the prolonged and long-winded discussions between politicians and bankers. There, the most pertinent questions remained unasked and honesty and genuine remorse were casualties.

So much Australian-owned agriculture has been lost. Too many farming families and their communities have perished. I have no idea how Bill Mott survived a lifetime of farming in this land of drought and flooding rains, let alone the bureaucracy, bankers and bastards he was subjected to. He paid a high price.

Bill Mott, like so many Australians, believed in the basic values and ethics celebrated within Australian culture, only to be betrayed by the very institutions which pretended to honour the same.

Kathryn Spurling
2020

Introduction

The music rises and falls, a lovely restful melody which inspires you to watch the real estate video. Taken from high above, all is idyllic. After recent rains ,the property is an expanse of green. The beautifully kept Queensland homestead, sheds and yards are testimony to generations of farmers and pastoralists who understood this land, whose tears and laughter had been heard for hundreds of years and whose sweat had long soaked the dirt beneath their feet. The sublime real estate video gives the impression that the owners wished to move of their own free will. Nothing could be further from the truth.

Bill Mott stands quietly looking over rich agricultural land in the Inglestone district of Queensland's Western Downs. His family traces generations of farmers and graziers who were the salt of the earth types, hard-working Australians around whom the nation's mythology emanates. This was once his and was to be passed to his children, but no longer. He must be careful to remain on the road which skirts this black soil property. As the sun settles further on the horizon, symbolically his shadow casts long and dark over the land he loved and laboured. It is all gone now – the dreams, the future, the reason to live – and it hurts.

It was easier to understand once. Bill Mott left school early to help his parents on the land, never wanted for anything else but his own land, hard work, family, the sense of community derived from country folk all pitted against the harsh reality of this ancient land and its climate of drought and flooding rains. Overseas wars had taken so many of their own, torn families and communities apart, but they had to believe it was for the greater good, for the safety of the land they worked and the preservation of a history and culture they held firm.

In 2019, this no longer reflected reality. Rural communities were fast disappearing. Land once owned and worked by Australian families was increasingly sold to overseas interests intent on their own nation's food and water security, and none attuned to Australian history and culture or community.

It is of no comfort to Bill Mott that he is not alone. His story highlights that of thousands who had been subjected to a tsunami of institutional dishonesty, greed and corruption, poor bureaucratic legislation, conflicted and complicit governments which has robbed Australians of more than their land. It is the story of a farmer named Bill Mott and it is the story of so much more.

One

'Nothing short of a Royal Commission…can be of any possible use to us at the present time.' William Gordon Mott, 1932

Bill Mott was born on 9 January 1957, in the heat and summer haze of the Western Downs of Queensland, to David and Joan, named in honour of his grandfather, William Gordon Mott. Like his father and grandfather, he became a farmer/grazier. In 1932, William Gordon Mott had called for a Royal Commission into poor management and corruption within the wool industry. In 2018, his grandson, Bill, added his voice to the Royal Commission into Misconduct in the Banking, Superannuation and Financial Services Industry. In 2019, Bill Mott's fight against the corrupt and fraudulent dealings of a bank which took his farm, inheritance and future, continued.

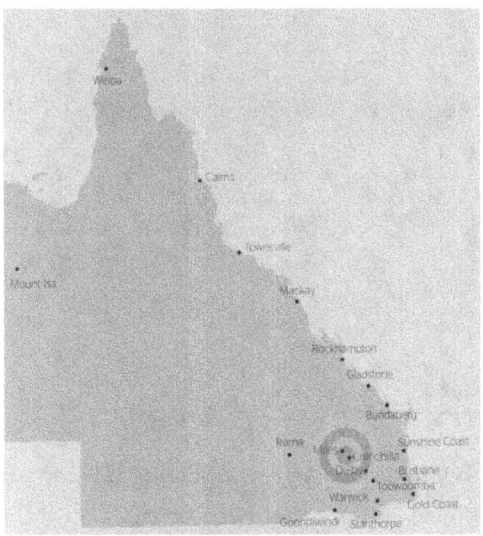

Western Downs Visitors Guide

You could be forgiven for being unaware of the location of Meandarra, the closest town to where Bill Mott called home in 2018. It would likely be no easier if the regional towns of Tara, Glenmorgan and Surat were mentioned. If the larger towns which make up a triangle around Meandarra were mentioned – Roma, St George, Dalby – and Toowoomba, closer to Brisbane than Meandarra, there might be more recognition. The country west of Toowoomba is as familiar to Bill Mott as the lines on his worn hands. The names of Ballaroo, Bollon, Bindebango, Jimbour, Hebel, Nebine, Rocky Bank, Naldera, Thallon, Nindigully, Mungindi, roll off his tongue with ease and sentimentality.

For hundreds of years, his ancestors had been defined by hard work, a simple lifestyle and the ever-challenging Queensland climate. His grandfather William and grandmother Eveline 'Maud' epitomised the pioneering spirit which Australians liked to believe defined them. William Mott ran a dairy farm south of Toowoomba, Queensland. In November 1918, at the age of twenty-two, the slightly built 170 centimetres tall farmer enlisted in the Australian Infantry Force (AIF) and was shipped overseas to fight for Britain and her dominions.

The desire to enlist for the empire was strong among country men.

Private William Mott, Regiment Number 921, arrived in England in July 1916. He lost little time in becoming better acquainted with the locals and on 11 November 1916 married Eveline Maud Randall from Islington, Trowbridge, Wiltshire. They had mere weeks before Lance Corporal Mott was dispatched to the horrors of war on Europe's

Western Front. Less than a month after he was promoted in the field to corporal, Mott was wounded in action during the Battle of Messines and from the Australian hospital in Rouen he was shipped back to England for further treatment for his back injury and exposure to gas. Before Christmas 1917, William was returned to the frontline.

Promotion to lance sergeant came two days before he was buried by an enemy barrage in July 1918. Shell-shocked William Mott was shipped back to England and placed at a training depot. He returned to Australia in 1919 and his English wife followed.

Evelyn 'Maud' and William Gordon Mott

William applied for a soldier settlement and was allotted acreage outside Quilpie, Queensland. Maud Mott arrived from the moist, gentle green of England to a tent pitched in a hot and harsh landscape bordered on one side by sand dunes. Her first duty was to cook in a kerosene tin on an open fire for her husband and six Indigenous stockmen. This first attempt was so unsuccessful a meal was only forthcoming with a lesson from a stockman. For Maud, the only contact with the outside world was the truck which delivered mail and provisions each fortnight when it could. It was a decade or so and five children later before the family moved out of tents into the cottage William built by hand. Two more children were born, including the youngest, David, the only son. The family bred sheep and cattle on the property named 'Old Kyra', 950 kilometres west of Brisbane and endured as best they could.

In 1929, William, on behalf of the Graziers' Association, the Cattle Growers' Association, and Selectors' Association of Queensland, was pivotal in an affidavit being presented to Queensland authorities, calling for a reduction in the award wages of shearers and station hands. The unrelenting drought had 'caused great losses of sheep and cattle', which in turn had 'caused great hardship amongst sheep and cattle owners'. There had been heavy rises in the cost of production due to feed supplementing 'heavy expenditure in endeavouring to save stock'. The price of sheep and wool had 'considerably declined', threatening 'the very existence of stock owners'. (*The Sydney Morning Herald*, 20 August 1929)

Life on 'Old Kyra' continued to be filled with adversity. A daughter was dragged to death by a horse and was buried in the sand dune behind the house. The workload was as endless as the unforgiving environment. When there was a reasonable wool yield, the profits failed to enrich the men on the land; instead they appeared to fill the pockets of the middlemen. William became more vocal.

> We are all acquainted with the system from the sheep's back to the saleroom – it is perfectly clear and above board. What actually happens after the sale of the wool, is a complete blank to most growers. (*The Brisbane Courier*, 1 September 1932)

William Mott believed that possible fraud and corruption could only be revealed and the 'rehabilitation of the wool industry' achieved through a Royal Commission, 'composed of genuine wool growers (not agents or managers)'. A Royal Commission 'could put witnesses under cross-examination, demand sale documents, etc'. (*The Brisbane Courier*, 1 September 1932) Remarkably, his grandson Bill would argue the same some eighty-six years later as he faced Australian parliamentary committees to request assistance in obtaining National Australia Bank (NAB) documents proving the level of fraud which had robbed him of his farm.

Although the federal and state government scheme of awarding farms to World War I soldier settlers seemed charitable in principle, they were commonly awarded the most challenging and inhospitable rural holdings. There was little option but for William and Maud Mott to battle on in the west of Queensland. In 1940, William reported to the media that in the Quilpie region

> Very hot weather, rain badly needed. Grasshoppers have made their appearance on some holdings and are eating off what green there is. (*Charleville Times*, 12 April 1940)

The sand dunes behind the homestead were moving and within fifty years would consume the home William and Maud built.

Years before, however, they had managed to sell the property and move east to properties near Muckadilla, forty-one kilometres from Roma, to grow wheat. The wheat crop provided little more than subsistence and William supplemented with a few dairy cows. His prewar experience and his butter became highly sought after. A family setback caused the couple to purchase 'Nalderra', a 20,000-acre property, a hundred kilometres from Roma – this they believed would provide their only son, David, with a secure future.[4]

David and Joan Mott on their wedding day.

David married Joan and the new family lived in one of two homes separated by a walkway, William and Maud in the other.

William 'Bill' was born in 1957, his sister Leonie eighteen months later, and the family worked to make 'Nalderra' a successful sheep property. William's health deteriorated and he was diagnosed with bowel cancer. David struggled with the increased responsibility. In an attempt to lessen the burden, William sold half the property and invested the money in an oil company. This, he believed, would more than cover his death duties. It was the worst decision he made. The acreage decreased rapidly in viability as sheep numbers needed to be reduced from 5,000 to 2,000. Probate on William's death caused a major financial loss, the oil company failed, and the money invested was lost.

The responsibility for the property and three generations of Motts all too soon fell on the shoulders of the younger William, not yet a teenager. He too needed to place trust in sheep and wool and, as his grandfather before him, was victim of individuals and an industry with little understanding or sympathy for the man on the land.

Two

'I found him sitting behind his ute with the end of his rifle's barrel under his chin.'

Bill Mott does not remember wanting to be anything else but a farmer/grazier. Perhaps it was in his blood; perhaps it was due to the mesmerising vastness of western Queensland and its kaleidoscope of strong colour – reds, black, orange, yellow and, in a good year, different shades of green. For a child who revelled in the physical, the company of animals, and the very mystique of nature, there could be no better life. The many chores and hard work were simply normal.

School was a necessity but given the remoteness of 'Naldera', schooling was complicated. Initially, it was correspondence from Charleville for four years and then school at Roma State School. He boarded in a Roma Methodist hostel for one year. The hostel matron was the proverbial battleaxe who was selling the hostel food on the black market, so there was no sense in asking, 'Please may I have some more?' The staple diet was bread and lemon spread. There was never enough to eat. Bill quite naively found a way to supplement his income. Because he was only ten years old, he was given his own small room between the boys' and girls' dormitories. Initially, he didn't understand why older boys were willing to pay him to 'borrow' his room for a couple of hours. Bill earned two shillings from every boy and girlfriend who sought two hours of isolation.

It is unclear if this entrepreneurial talent caused the end of his hostel stay but for the next six years he boarded privately. Schooling began to diminish from the age of twelve as he took on more work and responsibility as his father became more incumbered. David Mott never had the same affinity with the land as his father and son. Long weekends for Bill became standard. For the children within farming communities this was not unusual, but when still a student of Roma High School and before he was sixteen, Bill Mott was managing 'Naldera'. Bill had purchased an old utility for $200 and secured a special driver's licence at fourteen on the proviso that he report to the police superintendent when he drove in on Monday or Tuesday morning and before he left on Thursday or Friday.

Drought forever challenging

At nineteen, his life took a tragic turn. While mustering stock, Bill's young horse put a foot into a rabbit hole and Bill plummeted head first. Bill heard the crack and then terrible pain consumed him. He was a long way from the homestead and couldn't move. With darkness approaching, the situation was grim. He threw rocks at his horse until it trotted away – Bill hoped his mount would return to the homestead. David became concerned when his son had not arrived. Seeing the horse waiting riderless at the gate, David mounted a motorbike and tracked his son. There was little his father could do but return home for a vehicle and help. The nearest hospital was an hour and a half away.

Bill's stepsister, a nurse, arrived with her husband. Bill believes her nursing experience saved his life, particularly her refusal to allow him to drink; 'all she let me do was suck on a wet cloth'. They lifted him carefully onto a mattress in the back of the utility. It was a blur of pain as he was moved to hospital. The injuries were extensive: pelvis broken in three places, torn tendons attached to the pelvis, a broken hip, broken ribs and internal injuries. He was taken first to Roma and then Toowoomba, where doctors somberly explained to his parents that if their son made it through the night, he should survive.

Bill was unable to move his legs and was told he was unlikely to walk again. Such a diagnosis is shocking for any family but for one whose very livelihood depended on a physically strong son, it meant despair. Three months in hospital and months of rehabilitation were brutal, which Bill refers to simply as 'a hard slog'. There was so much healing to be done and tentative steps using a walking frame.

Twelve months after the accident, a twenty-year-old Bill Mott was helped onto a tractor. His impatience with the walking stick was unabated, but sheer determination resulted in his proving the medicos wrong.

As if making up for lost time, life changed rapidly. Bill was dating Jacqueline, a farmer's daughter and sister of his best friend from when she was sixteen. When she turned seventeen, they became engaged and they married when she was eighteen and he was twenty-two. The Mott family had purchased the adjoining 14,000-acre property 'Dueidar' in 1978, to increase their sheep holding. Drought hit the Western Downs. Destocking meant it became impossible to meet the loan repayments and in 1980 they had no option but to sell. Fortunately, regardless of the big dry, the land value had risen which enabled Bill to pay off the loan and buy his parents out of 'Nalderra'.

Jason was born in 1980, Leanne was born in 1981 and Ben in 1983. There is a smile: 'Jacqui's father bought us a television and there were no more kids.' They sold 'Naldera' and purchased 'Rocky Bank' in 1982. It was 7,000 acres, a hundred

Youth and happiness

kilometres further west from the town of Mitchell. They also had a 30,000-acre 'Homeboin' block a hundred kilometres from their house block at 'Rocky Bank'. 'Poor' is the word he uses to define these years, but also 'happy'. 'It was dry sheep equivalent and that was all you could borrow and make off the land.' There were three years of drought and for four they were feeding their livestock. The workload had increased dramatically, and things were so dire Bill needed to work off-farm, long distances away, three days a week, 'just to buy groceries'. He would return once or twice a week, but the children were normally in bed when he arrived, and he would leave before they woke. 'When I came home one weekend, young Ben screamed. He didn't know who I was.'

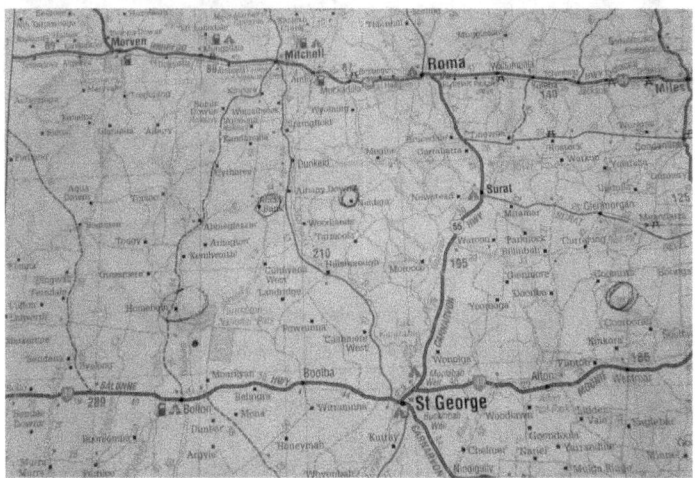

Vast distances in Queensland: the properties of Rocky Bank, Naldera and Homeboin

They decided this was not the family life they wanted, so they sold 'Rocky Bank'. Bill's careful management and improvement of the property enabled the young family to buy the house block with its old homestead at 'Homeboin' near Bollon. It was a hard-earned step up. 'Homeboin' and 'Bindebango' were originally part of the Australian Pastoral Holdings, which covered more than million acres, a pastoral empire leased by the Queen of England. The empire had been impressive, stocked with 60,000 sheep which took four months to shear in several woolsheds, one which boasted twenty sheep stands. In 1968, the Australian government resumed most of the acreage at the end of the ninety-nine-year lease and the remainder was passed back. The million-plus acreage was broken into 30,000-acre holdings and offered at the unimproved value to property owners within a hundred miles. 'Homeboin' promised a better future for the Mott family.

A year after they moved into the homestead the drought broke. Unfortunately, lightning struck one end of the house and started a fire. Jacqui and the children escaped in the car to neighbours, but the homestead burnt down. Bill shakes his head at the irony: the storm broke the drought and burnt down his house! He built a cottage and started a larger house which would take years to complete but they managed to buy two more parts of 'Homeboin', bringing the holdings to 90,000 acres. The children needed to move to Roma for their schooling and it was a wrench for their parents.

Life for women in the bush has always been arduous. They needed to be resilient, independent and able to undertake all manner of tasks. By the time she was twenty-one, Jacqui had three children and her husband needed to be absent to provide for the young family. Loneliness and lack of adult support made the days and nights long. Then, with the children needing to move into Roma from such an early age, it was particularly difficult. Jacqui moved to a friend's property outside Roma for several

Drought, fire and flood were part of life on 'Homeboin'.

years so that she could be closer to the children. Bill became more isolated at 'Homeboin' and there was even less reason not to work very long days. 'Homeboin' was timber and wire grass so he cleared much of the property, developed bore water reticulation, and improved pastures.

It took years for Bill to build the family home.

In 1932, his grandfather had declared that fraud and corruption could be revealed, and the 'rehabilitation of the wool industry' only achieved, through a Royal Commission, 'composed of genuine wool growers (not agents or managers)'. The grandson was unaware of this activism, but it likely would have made no difference. The saying had been, 'Australia rides on the sheep's back.' A ram's head had adorned the Australian shilling and in 1966, with the introduction of decimal currency, the face of pastoralist John Macarthur, and a sheep, took pride of place on Australia's $2 bill. Bill too put his trust in sheep and the wool industry and, like his grandfather before him, faced ruin because of a deeply troubled industry, riddled with fast money, middle management arrogance, political interference and bureaucratic incompetence.

The 1980s launched a decade in the business cycle of reckless optimism, excess and greed, a time of fantasy when speculators attempted to defy economic gravity. Alan Bond was a poster boy of a group of Australian entrepreneurs who dominated the 1980s. He helped bankroll Australia's bid to break the 132- year United States stranglehold on the Americas Cup, winning the cup in 1983 with yacht *Australia II*. Australia's prime minister, Bob Hawke, was jubilant and exclaimed, 'Any boss who sacks anyone for not turning up today is a bum'.

Australians were quick to buy into the admiration and excess. Favourable climatic conditions, rising wool prices and merino genetics, were seen by high flying investors, speculators, and corporations, as an opportunity for quick profit. None understood the industry from the paddock up. Reflecting on those heady days of success and excess, Bond said in an ABC interview,

> In the 1980s, there was a sense in the business community that you could stretch the envelope… If you were doing an acquisition or you wanted some planning permission, then the lawyers would work on the principle of 'How do we get round the laws as they are today?'(*ABC News*, 5 June 2015)

The bubble burst and the entrepreneurs crashed. Their losses were huge. Few Australians realised how far-reaching the economic crises stretched and the impact on men and women on the land.

But also implicit was the lack of flexibility and arrogance of politicians and wool industry executives. William 'Bill' Gunn was a bulky, 195-centimetre, rough-cut Queenslander who took pride in his outback, bushie, gauche appearance: 'a big man with big ideas who transformed several sectors of Australia's rural industry'.[2] Beneath the crumpled suits and dated hairstyle, Gunn was a crafty individual, determined to cement his own influence on the industry and engineer government intervention. As chairman of the Australian Wool Board, he achieved much in the promotion of the qualities of wool, but ultimately reduced the demand for the fibre he worked hard to promote. John McEwen, caretaker prime minister from 1967 to 1968, leader of the Country Party from 1958 to 1971, and venerable politician, was best friend to Gunn. Aided by McEwan and other Country Party ministers, Gunn politicised the industry and achieved statutory control over the entire Australian wool clip through the establishment of, the Australian Wool Corporation in 1973, and the introduction of the reserve price setting (RPS).

The business model was deeply flawed. The price reserve was a tool to drive the price up, which it did to a price few could afford to pay. Board and wool council executives treated overseas buyers with contempt. The resolute belief that overproduction and the ever-increasing stockpile would be resolved by monopoly and inflexibility led to the unravelling of what once was the nation's major export. Britain reduced its importation after joining the European Union. The Chinese market dropped, perhaps due to the arrogance, intransigence and price-setting within the Australian bureaucracy, but also because man-made fibres were increasing their share of the market. Wool executives refused steadfastly to allow the blending of wool with other fibres.

Bill Mott was facing calamities on all sides. The property woolshed blew over in high winds. He reconstructed it and the woolshed burnt down. It was a major undertaking to build a more robust structure and

Like all country kids, the Mott children were expected to work on the property whenever they could.

was only possible with the labour of his three children, aged thirteen to sixteen, now at boarding schools in Toowoomba. They journeyed home on weekends, just as their father had, to undertake near impossible endeavours. They ribbed him that he always left the hardest tasks until the weekends – he likely did, because eight hands were better than two.

A lot of hard labour and long days ensued, with Bill and his children precariously balanced on none-too-firm ladders and scaffolding a very long way from the ground, struggling to attach roof beams to supports, fixing metal to metal, in the heat, as the dust and wind blew.

Financial institutions changed names, Elders Pastoral to Elders Rural Bank, but the 1980s meant financial institutions had begun to lose their way in the desire to increase profits. Elders refinanced 'Homeboin' in 1988 and encouraged Bill to borrow more to increase sheep numbers and lease more land, as wool boomed. It was heady stuff. An older, wiser, Bill Mott would say in 2018, 'Poverty is the biggest land degradation when banks pressure farmers to overstock', but in 1989, 'Homeboin' stocked 25,000 sheep on 90,000 acres and harvested 800 bales of wool. For perhaps the first time in their lives, the Mott family felt just a little buoyant and profitable.

The best year

Then in 1990 the family took another personal blow. David Mott borrowed his son's new car, crashed head-on with a truck and was killed – he was sixty-two. Bill moved his mother Joan to 'Homeboin' from where she and David had been living in Roma, but the death of her husband exacerbated her poor health and dementia set in. Joan moved to New South Wales, to be closer to Bill's sister Leonie.

When it became clear the wool industry was in turmoil, the active players blamed the others. The Australian government was preoccupied with discord within, as Paul Keating challenged Bob Hawke for the leadership. The producers, who had been vigorously encouraged to increase the wool stockpile by bureaucracy and financiers, had nowhere to hide. Bill Mott believes, 'The politicians just treated the wool industry with contempt and as a liability.'

The Australian Wool Board had introduced a producer's levy on wool of ten shillings ($1) a bale. This was increased to two pounds and four shillings ($4.40) a bale, 'to fund the research and development'. It increased again with an additional levy to research new markets. In 1989, Bill Mott paid thirty-three per cent of the profit from his wool clip to the Wool Corporation for the levy on the wool stockpile. The Wool Board refused to allow more stockpile and only accepted fresh wool from growers. There was no effort to discard the cardings, the poor quality wool, from within the stockpile.

The Wool Corporations RPS collapsed in February 1991 and the industry imploded with an estimated $7 billion loss. For producers, the demise had dire consequences. Sheep which cost between $30 and $40 per head could not be given away. With the disintegration of the wool industry, Bill was left with a debt of $650,000.

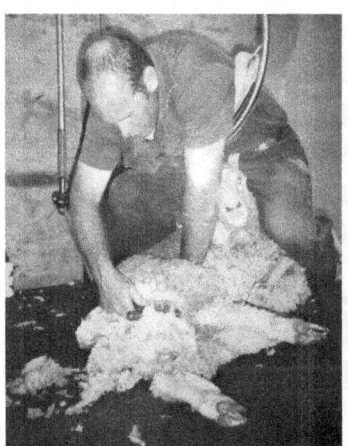 

Hard round the clock work could no longer save wool.

From the bowels of a federal government office came the decision which bureaucrats believed would in part resolve the wool glut and financially assist pastoralists. Graziers were to be paid $2 for every sheep they shot. It was a decision which could only come from those with no familiarity to agriculture, or empathy for livestock and those who spent their lives nurturing their animals. It is an episode which still brings tears to the eyes of this tough man of the land, one Bill Mott admits he is least proud of, and it is not enjoyable to listen to. Facing financial ruin, there seemed no alternative but to accept the government bounty, shoot 1,500 sheep a day, and push them into a deep hole. He discovered belatedly there was another task he should have undertaken before he covered them with dirt. For a man who cared for the welfare of his animals, the task was hideous.

And then there is another story.

There was a young bloke on an adjoining property who needed to shoot his sheep. When I was reloading, I couldn't hear him doing that, so I went over to check on him. I found him sitting behind his ute with the end of his rifle's barrel under his chin. I managed to get his gun away and called my wife to come out for him. Then I had to shoot his sheep as well as my own.

The man who was Bill Mott changed. The void between him and Jacqui widened and by 1994 she was living in Roma. In 1996, she had moved to the Sunshine Coast and they divorced in 1999.

'I couldn't really blame her. I changed, I was a pretty ugly person.'

There never was a Royal Commission held to investigate corruption and political intervention in the Australian wool industry.

Three

'Dingoes were killing twenty sheep a night and maiming 200 more.'

The Western Downs Visitor's Guide is a beautiful publication but in 2018 it lacks credibility. Resplendent on the colour wraparound cover is a group of four young white Australians walking their horses towards a vista of green and red plains. The trees in view are clearly not under stress from a seven- or ten-year drought and clouds linger promisingly above. The two male and two female figures could have walked out of a capital city real estate advertisement espousing the merits of living on an urban fringe. They are hatless, look cool, fresh and clean, and hardly resemble the population they are supposed to represent. The region's citizens are proud, as is their history, but both have been diminished.

Small country towns struggle to survive, the rural decline obvious in the closed shopfronts and abandoned houses. The region's citizens are ageing as the young seek easier and better-remunerated occupations in larger centres. The decline is due to a combination of factors which have occurred over time. Families once occupied houses on expansive properties to undertake agricultural work. They were part of the larger community, a community which shopped, went to school and formed the social fabric. Bill Mott recalls the community as it once was. There was support for local retail outlets even though they were more expensive. Buoyant social gatherings and dances in the community halls marked the calendar. People supported each other, cared for each other; children of the district could find meaningful employment if they wished to stay. Now no longer.

There are numerous reasons for this. Machinery grew larger and more efficient, dwarfing the efforts of a human labour force, making it unrealistic and uneconomical for large properties to employ agricultural labour and house their families. In the 1980s, Australia adopted open-market policies and agricultural markets became global – farmers were exposed to global price changes. The open market created the imperative for a 'get big or get out' philosophy, larger landholdings, the latest equipment – all of which required bigger loans and thrust the farmer onto a credit treadmill. Single-desk marketing boards like the Australian Wheat Board, which protected farms from price fluctuations, increased the impact of price changes and farmers were expected to purchase financial products to cover the volatility.

The iconic poem speaks of Australia as being a land of 'drought and flooding rains' and this is true; boom and bust have long been the burden of farmers and graziers. When seasonal rains don't appear, planting is delayed or not attempted, stock is reduced and then reduced again. Climate change has meant the extremes are more extreme. Drought assistance was reoriented to rely on market-based finances such as loans from banks rather than government grants. Different crops need to be cultivated; different sheep bred, for meat rather than wool. The landscape altered and shaped those needed to maintain it differently. Many who remain in rural townships are those who are unable to leave, have too little money, are unable to sell up or are of an age when it really doesn't matter any more.

Thallon was gazetted in 1911 and named after the Queensland Commissioner of Railways, J.F. Thallon. The town is surrounded by prime grain growing country and in 2012–2013, Thallon recorded the second largest grain intake in Queensland. There has not been a harvest anything like that since. Tourists are the only constant now, arriving to admire silos once bursting with grain, now carrying towering vivid murals by Joel Fergie and Travis Vinson titled *The Watering Hole*. showcasing the Moonie River district as resplendent as it ever was.

The Mott family called Bollon and then Meandarra home. Bollon sits on the banks of Wallam Creek, which has been known to flood the township. The soil is a brilliant red and visitors are invited to take the 1.2-kilometre creek-side walkway to appreciate the Indigenous Dreamtime artwork, or the more energetic may use the exercise equipment. These days, the creek is no longer a flood threat. Bollon, 634 kilometres from the Queensland capital, had a population of 334 in the 2011 census but, like the depth of the creek, this has dwindled. You can purchase a three-bedroom weatherboard home for $79,000 and the beautifully cared for office and attached two-bedroom home, currently leased as the local post office, is available for a negotiable $170,000.

Meandarra was the name of the first run used by pastoralist Archibald Meston in 1867. It is a town of civic-minded people, but it is a reflection of how it once was. Droughts destroy the soul as well as livelihoods and, in recent years, the Western Downs has severely tested those who wish to call the district home. When the Brigalow Creek doesn't run, it means drought again makes life an even bigger struggle and those who have the means drift east in search of work, perhaps even 361 kilometres as far as Brisbane.

The Meandarra Public Library opened in 1993 and remains open, as does the School of Arts Community Hall. The state school opened in

Meandarra State School

1913 with fourteen students. The meticulously cared for school building, built in 1997 to replace that lost in a fire, has an enrolment of thirty-five students. The main street is wide and long but there are now only a pub, a hair salon and a couple of retail outlets. The motel has four units and, together with the mixed business café, shop and house, is up for sale. The owner has aged and needs to retire. In 2016, 262 people filed census forms, and the highest percentage were aged sixty-five to sixty-nine, with the sixty to sixty-four age group next. There were ninety-four occupied dwellings and thirty-six unoccupied. The weekly household income of thirty-three per cent of the residents was less than $650.

Those who remain organise light relief as best they can. Events like the Flinton Races held on the Easter weekend continue to promise a

The wide but predominantly empty main street of Meandarra

fabulous day of bush horse racing and frivolity on a racetrack in the middle of nowhere near the banks of the Moonie River, between Moonie and St George. In Tara in August, there is the festival of Culture and Camel Races. On a far end of the triangle, you can get muddy in Roma's annual mud run, assuming there is enough water to make the mud. The Inglestone Community Hall that Bill Mott once viewed from his property is clearly in need of paint and repairs, but the locals gathered there in June 2018 for a bush dance.

History and heritage are prominent on the Western Downs. Ludwig Leichhardt passed through in the 1840s as he ventured further into the dry country. Central in every district, in every town, are the plinths, memorials and bronze edifices of Australian soldiers with their head bowed in catafalque party position. With a population of fewer than five million, 421,809 Australians served in the military in World War I, with 331,781 serving overseas. More than 62,000 were killed, 156,000 were wounded and many more suffered from what they had been subjected to in the Middle East and on Europe's Western Front. Per head, Australia suffered a casualty rate of around sixty-five per cent – greater than Britain, Germany, France, Canada or the United States. For Australia, it was indeed a pyrrhic victory.

Recruiting marches through country areas gathered country lads.

Country lads had swelled the ranks, targeted by recruiters for their horsemanship and physicality. Australian rural towns were never the same again and war memorials constructed throughout bore the names of families and communities who had sacrificed – names of a generation of sons who had ventured to war for the adventure of it, but also in the belief that the land their families worked needed to be protected from non- English invaders. Another generation followed between 1939 and 1945. Another 27,073 were killed or died on active service – the stonemasons were busy again adding names to memorials, often the same country families, the same rural communities. Further names were etched for still later conflicts.

There can be no denying the prominence of these memorials and the reverence awarded the names inscribed. Pride of place on Bollon's Main Street, is the Returned Servicemen Memorial Wall, featuring Flight Lieutenant George Steele (RAAF) who won a Distinguished Flying Cross (DFC) in World War II. An additional plaque commemorates the five sons of Winifred Khan who served in the same war.

The Bollon Heritage Centre run by volunteers is packed with interesting memorabilia tracing Indigenous and rural history. A central display is given to the local boys who left town to fight overseas – youthful faces in faded sepia photographs.

Meandarra war museum

Meandarra is dominated by the ANZAC Memorial Museum. Open in 2009 this enormous, spotless, collection of war memorabilia and stories of those who served, would be a focal point in any major city. The Canberra bomber might take pride of place, but Bill Mott has a strong connection with the World War I display.

His grandfather William Gordon Mott was not yet twenty-three when he left Australia to serve on the Western Front. The Mott family were fortunate. William, though scarred by the war, returned to be a

pioneer grazier west from here. His grandson respects that on all levels but remembers William 'refused to speak of the war...he certainly was a tough old bugger'.

Bill Mott needed to draw heavily on this example and tradition of endurance. Life had been arduous outside Bollon on 'Homeboin', particularly with the collapse of the wool industry, but Bill doesn't dwell on the 'bad' because that was normal – 'just life on the land'. The vast majority of Australians would interpret 'normal' much more benignly.

He had reduced sheep numbers dramatically in every definition of the word, but more pain was to come. 'Dingoes were killing twenty sheep a night and maiming 200 more.' The dingoes had been relocated to a nearby national park from a Queensland tourist island. The dingoes did not stay in the national park. It was heartbreaking. Bill engaged a professional dingo hunter, but the hunter could not keep up with the number of dingoes.

Bill made the decision to get into cattle. Mulga country was good for cattle. He paid attention to the lessons taught by cattlemen, particularly Mike Burello, who seemed like a 'cattle whisperer'. He learnt supplements benefited pregnant animals. Most importantly, Mike taught him the process called pressure release: positioning yourself with the cattle at the point of the shoulder. To get the animal to go forward, you stand halfway behind the centre of the shoulder; to have it go back, it is in front of the halfway point. 'You don't flog them, you just move, use your body to move the animal.' Sheep, on the other hand, do it their way.

Bill continued to improve his pastures and expand his bore water system and his herd of Brahman cattle rose from 500 to 5,000 and then 6,000.

Bill remained close to his children, who remained loyal to him and 'Homeboin'. Their love for the land continued throughout their teenage years. After Leanne completed year twelve, she studied at Dalby Agricultural College, and then worked as a jillaroo. The eldest, Jason, favoured the more mechanical and construction aspects of agriculture,

Leanne, Jason and Ben at 'Homeboin'

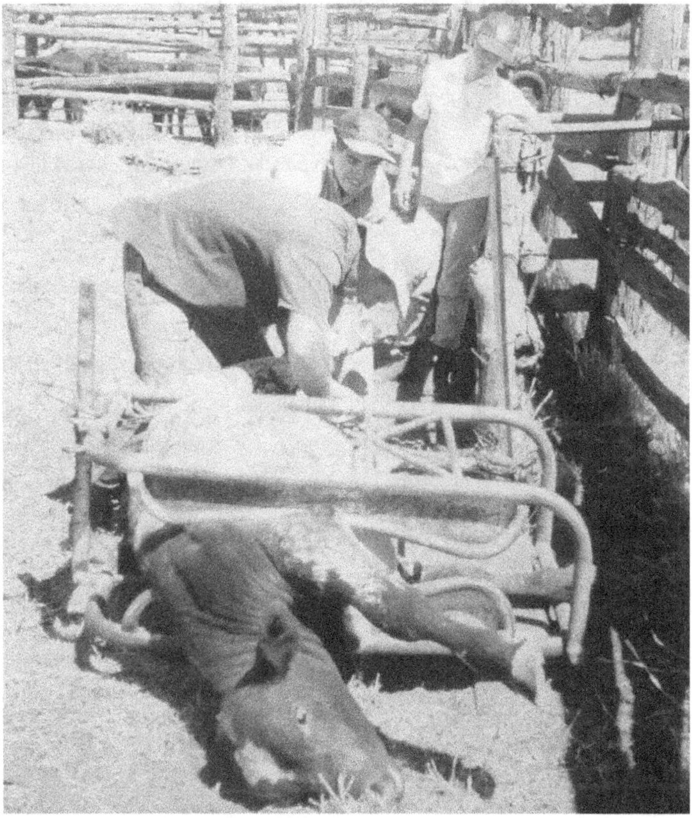

Jason and Leanne working with Bill

Ben Mott just wanted to be a farmer/grazier like his father.

and went on to study engineering. Ben, the youngest of the three, had inherited the affinity for farming his father had also demonstrated from an early age. 'Ever since he was a young fella, he just had this way with animals,' recalls Bill.

Ben, like his father, preferred the open spaces to the confines of the schoolroom and in 1998, having completed year ten, he decided that was his future. 'Homeboin' had been home for twenty-three years but Bill decided that for his children's future, the Mott Family Agricultural Holdings, needed to expand. 'Homeboin' was sold in 2006, a much improved, property than decades ago. The decision to sell was made easier because of the rare opportunity to buy properties in the fertile Inglestone district outside Meandarra. It would be another decade of good and bad seasons, but that was farming in Australia.

Four

'A couple of hundred thousand acres of really good country.'

The sun is setting on the horizon, a horizon which is endless – Australia's broad lands spread too far for the mind to comprehend, and no photograph can do justice to the vastness. Bill Mott picks up a handful of soil and lets it slip through his fingers. For a dozen years, he had done just this: knelt, felt the dirt and nodded. This side of the Inglestone district is black soil, the other side is red soil. Those who farm here know that it is better to cultivate the black soil first and move onto the red soil – it is all about moisture and the precious molecules which dictate a good, average or bad year. In dry spells, the black soil cracks, so when it does rain, the moisture is absorbed deeper. You just have to watch for the snakes which also emerge from the cracks. Porous soil, sandy loam, clay – all needed to be felt, watched, understood for productivity and the ability to use heavy machinery; and nothing is more important than planting.

The broad fertile plains of Inglestone, Queensland

Bill Mott could hardly believe his luck when he was given the opportunity to purchase two, then three, then four parts of this precious district. Inglestone was 'a couple of hundred thousand acres of really good country', prosperous Western Downs farmland with a lot of rough stuff around it. Generations of the same families had nurtured the land. It wasn't dry sheep country, it was all farming and 'you could borrow as much as you liked because of the productivity possibilities'. The banks were happy to lend, and Bill was happy to borrow. In 2006, the Mott family purchased 'Yambeen' and 'Nameerah', six months later 'Warooka', and then 'Woodville', 85,000 acres of 'magnificent country'. The future looked bright, assuming the weather was reasonable and you could trust the bank. In hindsight,: 'When we bought it, we paid too much for it, but the banks also overvalued it.'

Farming in Australia is a gamble, but Bill and his adult children had a love of the land and Bill only knew the credo 'if you buy and work hard, you are going to profit'. In 2006, he believed they had purchased properties of a lifetime – of lifetimes: his and his children's. He had only ever worked hard, so that would be no problem. He just had no control over the weather. To the question 'what is the average yearly rainfall?' Bill asks, 'What year?' Meandarra is supposed to have an average yearly rainfall of eighteen inches and that would have provided profitable agriculture, a steady income and money to pay loans.

Inglestone Hall

It is a little strange to see a golf course bordering massive fields, and for the visitor the question comes to mind, 'Who has the time and energy to play golf this distance from a major town?' Across the straight road is the Inglestone Hall, which had often echoed with country music and patrons resplendent in their Saturday best dancing the night away.

The golf course was on the corner of Bill's property. The saying was 'the closer you are to the golf course, the better off you are, and we couldn't get closer than if we were sitting on it'. It wasn't about the golf but the quality of the soil.

Inglestone golf course

In March 2006, they were allowed to remove weeds and organise cultivation. The plan was to grow wheat, chickpeas and sorghum and fatten cattle, and in August they took possession. The year 2006 was the driest year in the district for forty-six years. They only managed to crop late, and it was only useful for cattle grazing. They value-added cattle to survive that first year. Meandarra had its average rainfall in 2007 and they fully planted. But late rain meant a lighter crop and the grain got pinched (ran out of moisture). They ended up with high screenings, which lowered the value of the crop – it was just a feed wheat, no premium, but at least it would fatten the thousands of cattle they had brought from 'Homeboin'.

It was a difficult year as we ended up with four and a half ton and normally, you could look at $350 a ton; we were only getting $250 because of the poor quality.

It was necessary to change strategy and the decision was made to grow sorghum. Sorghum is the world's fifth-most important cereal crop after rice, wheat, maize and barley. Grown in clumps, it can, in a good year, reach to a height of around four metres. The grain is small and is traditionally used for livestock feed but increasingly in ethanol plants, because while it produces the same amount of ethanol per bushel as comparable feedstocks, it uses around a third less water. 'It is not a high-priced crop, but it built up the average, and the year 2007 summer turned out quite good.' The cattle were sold off, boosting the bank balance.

Due to drought, crops failed and fattening cattle provided the only viable option.

They had set up a feed lot for about 1,500 head, buying young cattle and selling them as heavy. It seemed the change of strategy was working and with a good planting opportunity, the sorghum produced a reasonable crop: '2008 was just a better year all round'. There was a collective sigh of relief. In the rural cycle, farmers need to long-fallow their country, leaving acreage without crops so it may recover its fertility. The ground is not ploughed, just chemically controlled. 'We had 16,000 acres of agricultural country but we were only planting 12,000 of wheat and chickpeas and the other grain sorghum or forage crop.'

There was a good start to 2009, but the cycle of 'drought and flooding rains' proved a reality. The year was wet, but not too wet and the 2,500 acres of chickpeas looked magnificent. As Christmas approached, so did the rain and as 2010 came so did more rain, twenty-six inches.

> It was so wet we never got to harvest. It was so wet we couldn't use our spray rig to spray them out. It was quite tragic really. You have to spray them to kill them and we had to get a plane in to spray them and then that night we had four inches of rain.

When the crop is sprayed, it wilts. The heavy rain flattened the crop to the ground. Bill couldn't even pick it up in his fingers. The rain had completely, destroyed their most outstanding crop. Also lost was 4,000 acres of wheat: 'it just fell over on the ground. It just perished.' When the wheat hit the ground and water, it shot the seed in the head. 'Funny, really. You have these little whiskers sticking out of the ground sitting up on the top of the ground with roots coming out the heads going everywhere.' Unfortunately, it was more heartbreaking than 'funny'. 'That should have been our big year…it wasn't.'

The only positive was that the Motts had purchased a header on tracks so they could harvest, whereas the contract harvesters could not harvest because they just bogged. They managed 3,000 tons of wheat. Wheat was a good price that year, $400 a ton – if only the deluge had not caused such a loss. It was so wet they could not even move onto the fields to plant sorghum.

The following year, they were not brave enough to attempt another crop of chickpeas, having been unable to harvest any of the crop the previous year. There was now another disincentive. India was a major importer of chickpeas. To encourage their own farmers to grow more chickpeas, the Indian government had introduced a sixty per cent tariff on the Australian import.

Drought, then too much rain, and then came the diagnosis of cancer. The cancer scourge had already hit good friends Jim and Gillian Meppem, who owned the Westmar Roadhouse. Jim's melanoma diagnosis was serious, as it had metastasised. Over the ensuing months, Bill helped as much as he could. The roadhouse was sold, and the couple and their two young children moved across the road from Bill. As Jim's health deteriorated, the family moved in to 'Yambeen'. The melanoma proved fatal. Over the ensuing months, the families grew closer and Bill and Gillian decided on their future as a couple. Because of the devastating effect of cancer, Bill decided to have a medical check-up.

> I had no symptoms, just decided it was the right thing to do…and then the tests came back with a diagnosis of aggressive prostate cancer.

Radical surgery was immediate. There was a great deal to cope with, but the Mott children took over the farming responsibilities.

Mott machinery investment needed to be huge.

Bill was anxious to make a speedy recovery. The family had needed to invest in expensive equipment to undertake the planting and harvesting required of the 18,000-acre Inglestone properties. The weather had treated them grievously.

With the soil wet, the decision was to concentrate on wheat for the remainder of 2010. There was a 'beautiful strike'. The wheat crop grew strong and tall, a 'magnificent crop of beautiful quality'. Then it stopped raining. From August, the clouds completely disappeared, and the yield was diminished. The harvest was only 6,000 tons when they had reasonably expected a yield of 15,000 tons. After costs, a farmer needs more than 6,000 tons to make any profit; 6,000 tons meant they were lucky to break even. To attempt to cover the losses in late 2010, the Mott family double-cropped sorghum.

Gillian and Bill had decided to postpone their plans in case Bill's cancer returned. In early 2011, they decided to marry. Rain had again become scarce but was forecast for the wedding day, 1 March. A marquee was hired and installed in the backyard of the homestead, which boasted Gillian's prize-winning garden, and families gathered to celebrate.

The night before the wedding, nine inches of rain fell. Dry creeks filled and flooded, so too the marquee, awash with three feet of water. The decision was made to move the ceremony to Inglestone Hall. Fortunately, all the tables and chairs were still loaded on a truck so, once a tractor towed the truck to the road, they could be taken to the hall. The community rose to the occasion and took over. The hall was not in the greatest shape, full of 'pigeon poo and spiders'. Guests pitched in and cleaned until it was pristine. The walls were lined with hessian and sheeting. Decorations, flowers and food arrived from seemingly every corner of the community and the wedding was memorable.

As the roads were flooded, guests camped out for the better part of a week. 'There was plenty of grog in the cold room, so everyone was very happy.' Bill had paid thousands for the hire of a marquee, which was a quagmire. Three months later, he received a bill for $120 for the use of the Inglestone hall.

A joyous wedding, Bill and Gillian with her children and a friend.

Life was good, he and Gillian were secure and happy, the future looked bright. Within five years, that had changed when the bank took it all away.

Five

'If you buy and work hard, you are going to profit.'

The washout of the 2010 winter crop caused a substantial financial loss for the Mott Family Trust aggregation. Despite the best efforts and hard work of Bill, Jason and Ben, the fickle western Queensland climate had wrought havoc. The initial elation of owning and caretaking this precious piece of Australian agricultural land had resulted in stark reality beyond the Motts' hard work and careful caretaking.

The losses were significant. Some 4,000 acres of wheat valued at a minimum of $300 per ton meant a loss of $1,200,000. The ruined 2,500 acres of chickpeas at $450 per ton added a further $800,000. Due to the sensible procurement of a header on tracks, they had been able to salvage tons of wheat. Rapid planting of 2,000 acres of sorghum on the moisture profile had returned $500,000 in May 2011. Bill was very aware that apart from Jason and Ben's livelihoods he now needed to provide for two young stepchildren, and the emotional burden weighed heavy.

Collective fingers were crossed for a better 2011. Early rain raised spirits and then the rain stopped. At the end of May, it was decided to dry plant a winter wheat crop of 2,000 acres but not to risk chickpeas. There were smiles when rain came in the middle of June and the green sprouts pushed their way through the rich soil. The Mott men raced to plant another 10,000 acres of wheat and, with the weather bureau forecasting good rains for the following three months, they planted another 2,000 acres of sorghum.

While they had faced the unusual occurrence of too much rain in 2010, the rest of Australia had recorded a large wheat harvest. An authority within the National Australia Bank (NAB), which now held the Mott loans, recommended the family 'forward sell' half of their expected

2011 wheat crop. The bumper wheat crop in other parts of Australia had meant a stockpile and increase in storage fees. Wheat is a perishable crop and unless it is forwarded quickly, weevils get into it and it is gone The silos owned by the Mott family could only store 1,000 tons. Anything further needed to be freighted day and night to the Gums, east of Meandarra. Cartage was commonly $16 a ton and storage at the Gums amounted to $20 a ton. When their wheat was delivered to port, it cost a further $35 a ton. Should there be another Australian bumper crop, the availability of storage would diminish and the costs would escalate. It was a difficult decision. Loans had to be met, the long-range weather bureau forecast had been promising and there had been advice from their lending authority. The Mott family were farmers; they needed to adhere to advice from those with expertise they did not have. They met the market and contracted 7,500 tons of wheat at an average of $225 a ton. The forecast rain didn't come. The crop failed.

Scrambling against the elements again, they sprayed out the 2,000 acres of double-cropped wheat, which left it with a good stubble, and 7,000 tons of top-quality Aph One wheat was harvested. The value of the wheat was $330 per ton but this had been presold for $225 a ton on the recommendation of the NAB advisor. The loss was $700,000. 'For our family, this was a disaster.' The fiercely independent Bill Mott had no option but to borrow from his new wife Gillian $450,000, the proceeds from the sale of the Meppem family roadhouse. The loan was unsecured and was to be repaid after the 2012 harvest. The Western Downs summer of 2012/2013 was hot and dry. It was considered too big a risk to incur the cost of planting even the less moisture-needy sorghum. There was no option but to put their faith in a mild and wet winter and bumper crop. More urea was spread and pre-emergent sprayed on 3,000 acres for chickpeas.

We upgraded our planter 24MT to 36MT with autosteer and purchased a new sprayer. Both machines had balloon payments of $120,000 at the recommendation of NAB.

The heavens opened with early planting rain. 'The wheat and chickpeas looked amazing.' The rainfall stopped.

With the dry end to the season and the wheat yield visibly dropping, harvesting was brought forward. Even with a dry finish, the chickpeas yielded $400 a ton. The wheat yield was down, however, and not of the top quality it had promised to be. Six thousand tons were worth only $240 a ton. The stress for Bill was growing. He had been unable to repay his wife's savings, which weighed heavily. He was an old-fashioned bloke who believed the man of the family was the bread-earner – it was he who needed to provide for both his families. As he let the soil sift through his fingers assessing its moisture content, and looked to the cloudless sky with desperate hope, he realised he had lost all control. With the dry summer, yet again there was not a good outlook for even sorghum and 'we had a long wait for another winter crop'.

Mott family machinery lay dormant for another summer.

There was every attempt to make Christmas a merry and happy family event, but Bill was uneasy. January and February storms lightened the mood and he closely monitored the moisture profile. Storms meant a patchy moisture level and without immediate follow-up rain, it was useless to even dry-plant wheat, so they planted 3,000 acres of chickpeas.

The weather continued to cruelly tease even the most experienced farmers. Good rain fell on 'Woodville', 'Warooka' and 'Namearah' in early June but not 'Yambeen', even though the properties were all within the Inglestone district. But the Mott family were grateful for any rain. They worked day and night to plant wheat, except for 2,000 acres at the top of 'Yambeen'. Though this included the coveted black soil, the moisture was too deep, and even more rain did not offer that opportunity. The rain throughout the district stopped. The wheat crop struggled. 'Harvest started early as the crop was literally dying as it ripened.' The chickpeas were short, and dirt in the sample meant they were downgraded. 'They were then shedded on the farm with the intention of having them graded but the market price was too low.' Only 5,000 tons of wheat could be harvested and it was only feed grade and sent to a local cattle feed lot. The chickpeas could also only be sold as feed. For the Mott family, it was 'devastating'.

Bill travelled to Roma to confer with the NAB manager. He had always had a friendly relationship with the manager – many beers had been consumed together and entertaining bloke chat shared over many lunches and hours. Bill had trusted the advice given. This was the financial expert, whereas Bill had left school at sixteen and was a farmer. The small print of contracts was constantly complex, but he had always been assured everything was in the best interests of the Mott family. By early December 2013, Bill feared he was unable to service the following year's debt, and this was quickly confirmed at the bank meeting. He asked if NAB was willing to support the family for another year?

> We agreed the best plan was to continue to maintain the properties and farming practices as this would be the optimal way to realise the true value of land and keep an income flowing.

Bill asked if the bank would defer the property and machinery repayments for one year so that money coming from the next harvest could be used for the following year's crop. The manager agreed to arrange a meeting at the Roma branch with an official from the debt recovery department and in the meantime he and another would inspect the Mott properties and undertake a trading budget.

Clouds gathered over 'Yambeen'.

The best Christmas present arrived early in the form of rain, and the smiles bringing in a New Year were more about additional rain. It was joyous to work the broad land again, every inch of it. At 'Yambeen', the final 2,000 acres could be planted with grain sorghum and 2,000 acres of ex-chickpea red soil, 'Warooka', could be planted with millet.

The smiles were wiped away at the next meeting at NAB Roma when the debt recovery official made it clear that he was unwilling to assist. 'It was like hitting a brick wall.' After a cordial forty-year history between bankers and farmer, and such convivial relations with the local NAB manager, it was a shock. There was agreement that the current property market was in a depression due to drought and that the likelihood of selling any part of the Mott family holdings was a poorer option than the family continuing to farm. Bill pointed out that they had planted 4,000 acres of summer crop out of their own savings, so NAB agreed to pay for sprays and fuel for the following year's wheat crop.

The situation was overwhelming. Bill's credo had always been 'If you buy and work hard, you are going to profit.' This credo he had passed on to his children, but it had become clear that in farming in the twenty-first century it was never going to be enough. He had believed his children would continue the Mott farming dynasty, but that dream withered with the summer crops. Jason had moved to Toowoomba to start his own engineering company. Leanne had moved to the Northern Territory to work in the mining industry. Only Ben, the child with the most affinity to the land, remained. His new family's financial future was also invested in the Inglestone aggregation. Bill repaid the $450,000 unsecured loan to Gillian. There were now less than no financial assets left and the credit card and overdraft levels were great.

In mid-January 2014, Bill travelled to Brisbane for a meeting with a member of an agricultural legal firm. Another meeting was arranged between a legal team and NAB debt department officials in Toowoomba for early February. Meanwhile, Bill and Ben continued to work the farms, silently pleading with the elements for a good harvest and an opportunity to dig their way out of the financial dilemma.

The February meeting did not go well. Bank officials declared that receivers were to be appointed to manage and maintain the properties. It was a terrible blow. Bill and his legal advisors 'vigorously opposed' the decision, as it was an unnecessary expense given that Bill, with his son's assistance, could continue to maintain and improve the farms. The bank was not moved and refused to fund the next crop unless receivers were appointed. There was little option but to agree. Over the next month, the flow of official paper increased before receivers were announced on 6 May 2014. Bill and Gillian were to remain as additional managers.

The weather was good, and the sorghum and millet harvest proceeded. The receivers took charge of the profits. More rain enabled barley to be planted. Bill attempted to pretend it was business as usual and planted more of 'Yambeen' as the improved weather continued. The flow of auditors and other well-dressed individuals continued to arrive to inspect, carefully rubbing the mud of their shoes as they re-entered their late-model vehicles. Bill did his best to stay out of their way, driving the machinery he was more comfortable with and enjoying the solitude of the expanse of land he desperately was trying to keep. It was especially cruel that the weather during 2014 was the best in many years. The aggregation had never looked so prosperous, with crops tall and robust, Bill argued how this proved that after years of drought, the farms were again profitable. This year's harvest was likely to return around $1.7 million.

In early January, NAB officials had agreed that the Mott holdings could not be sold in the depressed market at a weak valuation. In June, the receivers, acting on the bank's authority, announced that 'Yambeen', 'Woodville', 'Nameerah', and 'Warooka', 18,600 acres of 'magnificent country', were to be placed on the market immediately. In 2006, the future had looked bright, if the weather was reasonable, and you could trust the bank. In the following seven years, the weather had not been reasonable and they had learnt the bank was untrustworthy. The 'distraught' Mott family, who had resolutely believed if 'you work hard you will profit', had 'nowhere to go'.

Six

'It had started with a lie and finished with a lie.'

Bill Mott was mistaken. He estimated to NAB officials that there could be a profit of $2,000,000 coming from the 2015 harvest. The weather conditions were the most favourable for ten years. The gross income was not $2 million but $6.5 million. The properties were under receivership and Bill and his wife Gillian had been retained as joint managers.

Bill continued to spend long days in the fields again planting and nurturing the properties for the 2016 harvest, hoping that the NAB could realise how viable the farms were, and that the family was capable of trading their way out of debt.

The weather into 2016 was ideal and the cropping 'exceptional'. It was difficult not to dwell on the cruel irony of the situation after ten years of harsh climatic conditions. 'Yambeen' and 'Nameerah' were planted with wheat and the yield was around twenty bags per acre. The harvest was over 9,000 tonnes of top-quality wheat for which $330 a ton was paid, and the gross income was $2.5 million. 'Warooka' and 'Woodville' were planted with wheat which yielded 5,000 tonnes, and the additional 3,500 acres of chickpeas yielded 2,500 tonnes, which were sold for $900 per tonne. The harvest came to $3.5 million.

Due to the best three years of weather since the Mott family purchased the properties in 2006, the gross income between 2014 and 2016 amounted to $12.5 million. Bill acknowledges that clearly 'farming is a gamble' and there had been too many years of drought. But any law of averages then resulted in 'a good year, if not a few'. This had been proven during 2014 to 2016, with 2017 looking as bountiful. The properties could not have been more prosperous, and his hard work ethic proven. The receivers and NAB were unmoved.

Little did Bill realise how widespread this transition was, banks calling in rural loans after drought and profiteering when the good times inevitably cycled through. This pattern had been revealed in some thirty-eight inquiries into banking and financial institutions since 2010, which rarely received public attention. These were commonly instigated by minority party and independent members of Parliament, more sympathetic to the rural community and not bound by major party politics. Successive federal governments had continued to avoid calls for a formal inquiry into the banking and the financial sector.

Many of the grievances had arisen from when the smaller rural financial institutions had been subsumed by the larger banks. In 2010, the Australian and New Zealand Banking Group (ANZ) had acquired 7,124 loans worth $2.3 billion from Landmark. Clients complained there was an immediate communication breakdown, and changes introduced; interest rates were incorrectly applied, and limits erroneously loaded. The same confusion occurred when Elders Rural Bank was acquired by Bendigo and Adelaide Bank the same year. In 2018, Bendigo would admit to fee overcharging, irresponsible lending practices and underpaying customers interest due on term deposits.

In the past, there had been financial assistance and institutions introduced which specifically addressed the rural industry when it was realised standard banking practices were ill-suited. The Commonwealth Development Bank (CDB) provided finance to primary production between 1960 and 1974 when funding could not be obtained at sustainable terms through regular banks. Additionally, the former Rural

Credits Department (RCD) of the Reserve Bank of Australia had provided seasonal credits to statutory marketing authorities and rural cooperative associations for primary production projects. However, the RCD was disbanded in 1988 when it was believed the private banking sector had adjusted to rural seasonal credit demands. This was not the case.

In 2013 independent Member of Parliament Bob Katter and Democratic Labour Party Senator, John Madigan, co-sponsored a Reserve Bank Amendment, for the establishment of a board to address rural debt under the umbrella of the Reserve Bank and be tasked with the reconstruction and development in rural and regional areas. This board would be called the Australian Reconstruction and Development Board (ARDB). As the explanatory memorandum to the bill explained,

> Rural Australia is struggling under an insurmountable debt burden, characterised by low farm income and lending practices of financial institutions in deregulated financial markets. In 1980, debt in Gross Value Farm Production was at 32 percent and this has escalated to historically high levels of debt, reaching 135.4 percent in 2012. With escalating debts, many farms and producers are facing foreclosure. Forced sales are widening loan-to-value ratios, leading to a risk of 'fire sales', which could precipitate a raging financial contagion that may not be contained to rural and regional Australia. (https:www. aph. gov.au/Parliamentary_Business/Committees/Senate/Economics/RBA_Amendment_2013/Report/d01)

During the debate on the amendment, it was cited that there was 'a significant mismatch between lenders and agricultural borrowers on capital arrangements'. Successive Australian governments had removed various risk-ameliorating systems, which had resulted in an even higher level of risk. Simultaneously, financiers had failed to adjust and had continued to lend the same product, one which was inappropriate for the degree of risk in agricultural undertakings. On 12 December 2013, the Senate referred the ARDB 2013 amendment to the Economics Legislation Committee for inquiry and report by 26 March 2014.

Bob Katter had been the elected member for the division of Kennedy since 1993. Initially a member of the National Party, he became an independent in 2001 before forming his own party in 2011. His electorate encompassed a massive 568,993 kilometres (219,689.4 square metres) from the Queensland Pacific to the northern Queensland outback. It was one of his constituents, Charlie Phillott, who became the first public face of the rural banking crisis.

Charlie Phillott was in his eighties and had been a central-west grazier for more than fifty years. During that time, he had faced all the classic rural hardships: fire, flood, drought, fluctuating cattle and sheep prices, highs and lows in foreign exchange. The family property, 'Carisbrooke Station', eighty kilometres south-west of Winton, Queensland, had nonetheless endured. The family had been respectful custodians of the Indigenous rock art in the property's rugged red-dirt outcrops. Tourists were welcomed to appreciate this example of Australia's first people's culture.

Federal MP Bob Katter, pictured with farmer Charlie Phillott (left), interrupted the banking royal commission with an outburst today. Picture: AAP Image/Glenn Hunt (https://www.qt.com.au/news/anz-admits-unfair-and-unethical-behaviour-towards-/3452126/)

The drought in the central north of Queensland was long and deep. By 2008, debt at 'Carisbrooke' was mounting. A $1.5 million interest-only loan was taken out through agricultural lender Landmark, to consolidate and expand the tourism business. The family's finances were purchased by ANZ. No longer was there a knowledgeable, friendly local manager. The long-distance communication between bank and client deteriorated.

The ANZ halved the property's value, which erased the family's equity. Despite the Phillotts having $7,000 left in an overdraft, the bank declared the family in default and doubled their interest rate. Until 2012, full payments were made, but the interest was crippling. Shortly after, the family was evicted. 'Virtually overnight the bank forced the Phillotts into financial oblivion and they had to walk away.' (*Queensland Times*, 7 September 2016) According to Charlie Phillott, 'They just sent the bailiff out to put Charles, my son, off the property – gave him fifteen days, I think, to get off.' (*The Saturday Paper*, 28 April 2018)

Charlie Phillott contacted Bob Katter. Katter and his son, Queensland state MP Robbie Katter, were aware of the growing rural crisis and had met with the Labor government in 2012. The government offered concessional loans. The last thing farmers needed was more debt. Bob Katter garnered media attention around the Phillott family which placed the ANZ bank in an embarrassing situation. An apology was forthcoming, and the family was allowed to return to their property in 2015. The high-profile media attention highlighted the unscrupulous financial practices and arbitrary nature of banking contractual procedures. Australians were becoming more aware and soon learnt that the Phillott family were the usual rather than the isolated example.

Bob Katter took issue with the fact that while the federal government had agreed to underwrite the big banks, they had not shown the same concern for the those borrowing from them at what were often unreasonable terms. The Abbott government like its successor simply offered concessional loans. More MPs and Senators voiced concern. New South Wales National Party Senator John 'Wacka' Williams urged Prime Minister Abbott and his successor, Malcolm Turnbull, to reconsider. Williams served on the Senate Economics References Committee which had held a 2013 inquiry into the financial regulator, the Australian Securities and Investments Commission (ASIC) following a series of financial and investment scandals. The committee, chaired by Labor Senator Mark Bishop, called for a Royal Commission into the nation's financial and banking institutions.

In 2013, South Australian Senator Nick Xenophon (Nick Xenophon Team) in support of the Bob Katter's ARDB, agreed it was his finding that the private banking sector had failed the farming community and that Australia was, indeed in the midst of a rural financial crisis. Xenophon cited West Australian sheep and grain producer Bruce Dixon, who was confronted by ANZ-appointed receivers a day after defaulting on his multimillion-dollar loan. Dixon's debt level had risen by $500,000 in eight months due to the high interest rate applied by ANZ. Clearly the bank had prepared plans to take rapid possession of the property. Dixon alleged the receivers threatened action unless his family exited the farm swiftly, even suggesting they could request assistance from the Australian Federal Police (AFP) Special Response Team. Local community frustration at the heavy-handedness was swift, with members of the Rural Action Movement allegedly impounding the receivers' vehicle by surrounding it with haystacks. ((https://www.aph.gov.au/Parliamentary_Business/Committees/Senate/Economics/RBA_Amendment_201 3/Report/d01))

In 2015, the Australian Green Party (the Greens) put forward a Senate motion demanding that the Royal Commission be held. Senator Williams broke ranks from the Coalition government by crossing the floor in support. Both the Coalition and the Labor Party voted the motion down. Prime Minister Turnbull repeatedly rejected calls, as did the banks.

One Nation leader Pauline Hanson was motivated to negotiate a Senate inquiry into farm lending following the treatment of One Nation Senator Rod Culleton, whose bank had foreclosed on his property. Culleton, from Williams, Western Australia, was another victim of rural debt and was involved in protracted legal proceedings after the ANZ foreclosed on the family farm. The courts dismissed all charges against

ANZ. In his maiden speech to the Senate, Culleton apologised for not being well-spoken due to his lack of formal education and that 'as a farmer, it is true, the only audience I have had to address up until now has simply been a mob of sheep'. (Hansard, 12 October 2016, p. 1673)

The speech nonetheless graphically expressed the predicament facing Australian farmers caused by bureaucracy, bankers and global influences. Culleton was a fourth-generation farmer.

> At the time I started my career in the wool industry Australia was riding on the sheep's back. The wool industry was large, prosperous and well regulated, protecting both the industry and its consumers against poor quality product and exploitation of its farmers and communities. (Hansard, 12 October 2016, p. 1673)

Culleton ventured into buying Australian wool clip in 1981. Privatisation put an end to competition. He believed strongly that 'foreign corporations' then became 'the beneficiaries of our tax laws and our lack of regulatory structures'. After January 1991, the price fell from 870 cents a kilogram clean to 430 cents a kilogram clean. The wool stockpile and debt of billions was 'a national disaster'. As Bill Mott had witnessed, the failure of administration and appropriate government policy caused huge financial and emotional pain.

In the Senate, Culleton spoke of how farmers were still haunted by what occurred.

> Our stock was worthless... The government...ordered 20 million sheep to be culled. The sheep were gassed and shot en masse as a result. I remember...farmers running sheep into the back of semi tippers, rolling the tarps over the top, shutting the back tailgates and gassing the young sheep to death – under instructions from the then government! I vomited over the fence....These memories still haunt me today...a national disaster was caused by implementing poor political decisions, made in ignorance, without appropriate research, by politicians who simply refused to go out to the coalface to learn the consequences of their actions. (Hansard, 12 October 2016, p. 1673)

Senator Culleton was disqualified from Parliament due to bankruptcy early in 2017. The Senate inquiry into farm lending chaired by One Nation's Pauline Hanson was due to report its findings at the end of 2017.

Despite admission by Australian Treasury officials that rural debt had trebled during the decade and a half to 2014 and the Reserve Bank of Australia's acknowledgement that rural debt had more than doubled between 2004 and 2014, the Australian Bankers Association (ABA) continued to deny that there was a 'rural debt crisis'.

In October 2016, Bob Katter, with the support of Queensland National Party MP George Christensen, put forward a private member's bill calling for a bank inquiry. The Greens supported the bill and this time, the Labor Party agreed. The 2016 federal election had resulted in only a one-seat majority for the Turnbull government. In March 2017, advocates jointly drafted a new bill for a Royal Commission. The banks and the Coalition government continued to oppose, until it became clear that Christensen and fellow National Party MP Llew O'Brien intended to cross the floor in support, thus jeopardising the government's grip on power. The banks publicly withdrew their opposition. The government had no alternative but to accept a discomforting backdown. On 30 November 2017, Turnbull and his treasurer, Scott Morrison, announced that a Royal Commission into Misconduct in the Banking, Superannuation and Financial Services Industry was to be established on 14 December 2017.

Bill Mott admits he has never been much of a reader and, though he had observed the difficulties and diminishing of his own farming community, he was unaware of the goings on in the chambers of the Australian government. He was just intent on his own situation and the overwhelming feeling of helplessness and confusion. Bill had never drawn a salary during his working career. With the receivers in possession of the family properties, as manager he was receiving $400 a day and Gillian was receiving $300 a day. A receiver employee, on a daily salary of $1,500, sat in his airconditioned vehicle all day, watching the

Mott family work. It was crazy money, baffling when the NAB had declared his properties a financial drain. Bill continued to plead for more mediation. The properties had more than shown their productivity recently and the current crops again looked excellent.

Unfortunately, the unrelenting stress had caused a rift between Bill and Gillian. Gillian was also fearful that the NAB would seize the $450,000 proceeds from the sale of her and her late husband's roadhouse. She was concerned for their two teenage children, for their education, and for their not having a home. Bill's anger with the world was palpable and there was nothing she could do to assuage it. The marriage was at an end and she retreated to Toowoomba. 'It was tough, tough on my marriage…a new marriage just couldn't take it.' For Bill, it was another devastating blow.

The NAB representatives had assured the Mott family repeatedly that their farms would not be sold for less than the $22 million valuation they had placed against the loans in 2006. It was clear to Bill 'that both land prices and grain prices were on the increase', so there was no need for a hasty sale. In September 2017, NAB and the receivers placed the Mott farms with real estate agents Ray White Rural. The property was listed as being 'open to expressions of interest'. For Bill came the

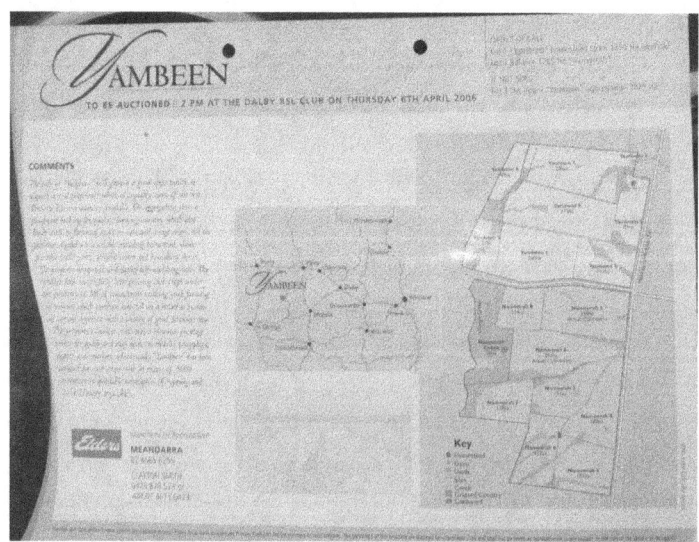

realisation that 'It had started with a lie and finished with a lie.' His family was informed that there would be a clearance sale on 24 November, and they were to be evicted from their home the following day.

Seven

'I was good and then they bled me dry.'

The emotions were raw, and it seemed to Bill Mott that he was standing outside watching his life fall apart. He had only ever wanted to be a farmer, had worked incredibly hard on the land all his life, could count on a couple of fingers holidays away. Now, weeks away from his sixty-first birthday, everything was broken.

The Mott family had been well-liked pillars of the community. They involved themselves in the rural district, willingly paid the necessarily inflated local prices, believing support for local services was vital to sustaining the region. They joined others to celebrate the good times when the creeks ran and the landscape flourished in shades of green, but were also known as people you could call in any emergency. Their prize-winning gardens had hosted a district fundraiser which 500 people attended.

Life was now surreal and Bill was struggling. Nothing made sense. Once the receivers took over, communication was minimal. The banks had said there would be no rushed sale at a time when the original valuation of $22 million was unlikely to be reached. Yet here was a lavish four-page real estate brochure displaying the Mott properties looking their very best, and eviction was imminent.

The sale was by tender, so as secret as it could be. Bill and his sons believed it would sell at a high price. In December 2015, a grazing property called 'Heswen' had sold for over $600 per acre. In May 2016, a still timbered, unimproved block without infrastructure, adjoining 'Warooka' and 'Woodville', sold for $660 a square acre. Their four properties must be worth in the vicinity of around $1,000 an acre. Bill's home was well maintained, with wonderful gardens, swimming pool, and silos and sheds similarly well cared for. The flourishing crops, planted and

nurtured by Bill and son Ben, were soon to be the property of someone else. Just perhaps there might be money left over after the debts were covered. Just perhaps it wouldn't sell, and they could stay on as managers, trade themselves out of debt and the property would be theirs again. Bill realised this was a pipe dream. Who would not want the Mott properties when they looked as they did in 2017? He was clutching at anything, going about the daily routine almost in a daze.

Neighbours were angry and likely would have purchased the Mott holding had the receivers allowed them to be split up and sold separately. Then, maybe, they could be sold back to Bill when he was more solvent. On the day before the tender process closed, an individual was observed in deep discussion with the agent. The following day, after the tender process had closed, he returned to sign purchase contracts. The illegality was riling and worse was to come.

It was revealed that a professional who lived in Sydney, New South Wales, had purchased the four properties covering 7,408 hectares (18,305 acres) for not $22 million but $8.2 million. Had they been counselled, Bill, Ben and Jason could have raised that amount using Jason's engineering company machinery as equity. It had been a fire sale, but for what reason? The anomalies and lack of consultation were perplexing. The receivers and auction house, under instructions from NAB, now moved on the Mott's farm equipment and household possessions. The community boycotted the clearance sale, but outsiders took advantage of the give-away prices. Three-year-old carefully maintained farm machinery valued at $1,200,000 sold for half price, $600,000 excluding GST. Auctioneer costs ran to $35,500. The auctioneer cost of selling other Mott possessions valued at around $485,000 cost NAB $80,000.

Outraged neighbours salvaged what they could for the Mott family. They stripped rooms and carted remaining personal possessions to a loaned house on a next-door property. Beds were prepared and meals left in the refrigerator. It was the only small glow on the wretched 24 November 2017 day of eviction. What had just happened? Bill attempted to clear his emotions well enough to be able to do some figures, and they were not logical. The NAB likely had spent $1,000,000 in receivers' costs; more than $6,500,000 in income during the 2014–2016 years; more than $6,000,000 with the rushed sale of the properties. The most fortunate interstate new owner had not only purchased choice country for a bargain price, but the harvest of the crops planted by the Mott family, was pure profit. The new owner never intended to live in the homestead, with its prize-winning gardens.

Bill Mott was a farmer. He had put his trust in the financial experts, he had to, and now he realised he had been 'set up to fail'. By 2010, 'I was good and then they bled me dry.' One year, he had paid $1 million in interest alone. He had banked with the NAB for decades. What had happened to the trust and caring customer relations? The deed of forbearance was created by the banks for the banks – which had one paragraph in the complicated and lengthy contract that said the bank could control the sale and could bring in receivers. The only thing in Bill's favour was that this meant he could not be declared bankrupt.

He was angry and intended to fight, 'I can be a stubborn bastard.' But how could he take on the bank? He had no money, so taking the NAB, which had a market capitalisation of around $70 billion, to court was not really an option. Being far away from a capital city meant he could be ignored, which meant he needed to get on the road and haunt them until they took notice. At the same time, some understanding needed to be gathered of the boxes of complex figures and financial statements – and it was crucial that he gained access to bank documents.

Soon, travelling the thousands of kilometres between far-flung city offices and Meandarra became routine, but of utmost importance was the welfare of his son Ben. Ben had taken the loss of the properties, one of which was considered his own, badly. So, like his father, with an affinity with agriculture and a desire to be nowhere else, Ben now lived in a two-room demountable within sight of what once was his future.

Excellent with equipment, he now undertook contract work to make a living – clearing, ploughing, planting other people's properties – and it was a difficult adjustment. Bill, the self-confessed hard 'stubborn bastard', struggles with the memory and the eyes drop to the floor self-consciously when he recalls the depression he observed in his younger son. Blokes aren't good at expressing their grief and Ben had always been a man of very few words even during the good times. Bill Mott was close to his children, a proud dad, and now he felt helpless to make the pain go away or where to go for help.

Suicide is endemic in Australian society and particularly the rural community, yet it remains a taboo subject and the tragedy receives little public attention. A university of Newcastle/NSW Government Health briefing paper released in April 2017 revealed,

> In every state in Australia, the rate of suicide among those who live outside the greater capital cities is higher than that for residents that live within them, and the rate has risen much higher in rural areas over the period 2011–2015. (Centre for Rural & Remote Mental Health; 'Suicide & Suicide Prevention in Rural Areas of Australia', 11 April 2017)

While it was found that rural residents were happier than their city counterparts, they experienced higher levels of insecurity due to their current and future being so dependent on 'too much or insufficient rainfall, too high or too low temperature, hail, frost, fire etc.', all beyond their control. It was found that fifteen- to twenty-four-year-old males in regional areas were 1.5–1.8 times more likely to end their life by suicide than their urban counterparts. The incidence was up to six times higher in very remote areas. It was estimated that 2,864 Australians committed suicide in 2014, an increase of almost 13.5 per cent from the previous year (www.lifeline.org.au). Rural decline, isolation and lack of medical services were listed as causes, but rural debt and harsh treatment by banks were also responsible. Bill Mott knew of fifty rural suicides in Queensland over a period of a couple of years. He was depressed himself but much more concerned for Ben. 'Ben suffered badly… I couldn't cope…that he couldn't cope.' He made sure he checked on Ben, home and away, very regularly and he locked away the farm's guns. The weight of feeling responsible was heavy.

On the night of 29 November 2017, the treasurer, Scott Morrison, was on a phone hook-up with Wayne Byres, head of banking regulator APRA, and Philip Lowe, Governor of the Reserve Bank. The financial guardians, frustrated by the Coalition's refusal to set up a Royal Commission, believed that a band of backbenchers led by the Nationals MP Barry O'Sullivan had procured the seventy-six cross-party votes to launch their own commission. Such a commission could not be controlled by the government. It could reveal information which might cause 'irreparable damage to the financial system'. They suggested the treasurer had lost control and 'the major banks were terrified'. (*The Weekend Australian*, 6–7 April 2019)

Morrison admitted that

> The maths in the parliament was going to lead to an unwieldy and directionless and haphazard commission of inquiry which would have done far greater damage – and that was not something that we believed should be allowed to happen. (*The Weekend Australian*, 6–7 April 2019)

Ironically, the week the Mott family was being evicted from their home and farm, the treasurer conferred with the chief executives and chairs of the major banks to discuss forming a panel which could review misconduct and award compensation. The bankers supported the proposal; it was more in their interests than a commission.

Prime Minister Malcolm Turnbull, a former merchant banker, was very aware of the possible consequences of the O'Sullivan bill. A commissioner appointed by the dissenting group could be equally disastrous. The PM had raised the conduct of a royal commission in cabinet but 'Morrison had been furious at the suggestion. Backflipping would make him look foolish for arguing so stridently against such an inquiry.' (*The Weekend Australian*, 6–7 April 2019)

Financial Services Minister Kelly O'Dwyer was instructed to convince O'Sullivan to delay the introduction of his bill by a day. He agreed.

On the evening of 29 November, a letter signed by bank chiefs and delivered to the treasurer agreed times were desperate and the best way forward was to submit to a Royal Commission. The following day, before O'Sullivan could introduce his bill, the government announced a Royal Commission into Misconduct in the Banking, Superannuation and Financial Services Industry was to be established on 14 December 2017, with the Honourable Kenneth Madison Hayne AC QC appointed commissioner. The financial institutions and the government hoped control could be achieved. Though constrained by time and terms of reference, the commission was to reveal a plethora of fraud, dishonesty, greed and inhumanity, and bring into prominence a word few Australians were familiar with: 'incentivise'.

Eight

'It is all about the outcome, trying to get rid of us, it is never about our experience.'

There was at last optimism within the rural sector – a Royal Commission would reveal the extent of the misinformation, unconscionable lending and financial dishonesty they had been subjected to. Finally, the Australian people were to be made aware of the terrible state-of-affairs faced by families on the land. The anger and frustration were deep and the expectations high. There was a litany of allegations including hidden documents, devastating land value devaluations followed by fire sales and foreclosures, even forged signatures and unauthorised refinancing. It was announced that the Royal Commission would hold public hearings in Brisbane and Darwin to assist witnesses in regional Australia. The flood gates were about to open.

In March 2018, National Party Senator John 'Wacka' Williams welcomed the pending hearings. 'Hopefully a message will be clearly sent to banks in future – don't give bad advice and don't rip people off'. (*Sydney Morning Herald*, 20 March 2018) Williams had been deputy chair of the previous year's Senate inquiry into rural lending. He had heard distressing stories of

> The stress on families, the marriage breakdowns, the suicides; there's been a lot of harsh goings-on by the banks for too long, and when it involves land, farms and breeding herds that have often been in the family for generations, you never get over being forced off. (www.bankvictims.com.au)

Shown at the hearing was a photograph of a handsome, beaming teenager in a broad-brimmed hat. His name was Maccie Whelan. The Whelan family bred cattle on their two Cape York cattle stations, 'Dixie' and 'Koolatah'. Their cattle were once worth $800 a head but due to market fluctuations beyond their control, in 2013–2014 they were fetching less that $100 a head. The Rural Bank began to exert pressure, raised interest rates and called in the $12 million debt. Maccie was sixteen when his father, Jim, pulled him out of school to help run the 1.7 million acres of bush country. They mustered cattle from 'Dixie' in the hope of keeping 'Koolatah', but the bank issued a notice of foreclosure. Maccie shot himself. The properties were taken. The Whelans' youngest daughter began her battle in a Brisbane hospital with bone cancer. Jim Whelan hoped he could speak of his son Maccie at the royal commission and of family devastation.

> These banks have destroyed lives, marriages, families; this commission has to hear how the pressure these banks put on all of us cost my son his life and caused so many farmers to take their own. I call it corporate manslaughter: they destroy you, bleed you dry and wreck your family until you have no money or spirit left to fight with. (*The Australian*, 18 June 2018)

In June 2018, they were lining up to attend and give their stories, like John Wharton, the mayor of the Queensland gulf country town of Richmond. Police and security guards employed by ANZ receivers physically forced the Wharton family off their cattle property 'Runnymede', and the family home of ninety-six years was hurriedly sold. The family had never missed a payment but when ANZ acquired Bank West's rural portfolio, the property's budget was altered by the bank and the family defaulted.

The optimism of victims took a hit when the commission's terms of reference were released. They were very restricted, so too the schedule. There was too little time for more than a few farmers to tell their stories. Overall 6,761 submissions were received before the commission sat and it was expected fewer than one per cent of those could be called. No survivor who simply appeared at the hearings was entitled to raise their voice. The chiefs of different banks were to be called separately.

In Brisbane, Senator Bob Katter led a formidable group of farmers into the hearings and the depth of doom was unmistakable. In Brisbane, 'there were jeers, hisses, boos and even cries of "shame"' when stories emerged and ANZ executive Benjamin Steinberg needed to be hustled from the court. As journalist ABC Daniel Ziffer reported, 'It was tough to hear' the 'tales of woe...we really got to feel the pain.' (*ABC News*, 1 July 2018)

It became quickly apparent to the commission that rural stories did not fit well in a tight box – they were complex and very different; the Australian landscape was broad and diverse. It was usual for farmers to be asset-rich and cash-poor, and success or failure depended on drought and flooding rains. Each harvest dictated some liquidity or falling further into debt. From region to region, let alone state to state, the landscape was so different – there could be a good season in one end of the country and drought in the other. When the value of a property dropped, the loan to value ratio went up. A bank might lend fifty per cent of a property valued at $1 million – $500,000. If they then revalued a property for whatever reason to $500,000, the farmer was up to one hundred per cent of the lend. The bank might then believe the loan unviable and foreclose. Agriculture was also dependent on shifts in global positioning of farm land relative to other forms of property, and Australian government agricultural policy, and trade agreements. What was deemed diplomatically best for the nation had an immediate impact – farmers were expected to simply adapt, quickly.

Arthur and Rhonda Cheesman were third-generation Victorian farmers. They ran their properties with son Rueben and his wife Katrina, both couples living in separate homes on separate hay, cereal, oil seed and pulse farms. Their farming had adapted to modern world demands with the growing of lentils, but all crops depended on rainfall and their region had suffered below average for years. Loan interest costs, in 2007 and 2008, accounted for 40 per cent of their total farm income. They had been Landmark customers since 2004 and the Landmark loan book was acquired in 2010 by the ANZ. The bank instructed them to sell assets to pay down their debt.

The Cheesmans sold one of their properties. In October 2011, the family were notified that unless they took immediate steps to 'sell the secured property by the end of November', the properties and their homes would be sold by auction. When they were unable to sell, an auction was organised for January 2012. The family pleaded with the bank that their houses be excluded, because 'they would have nowhere else to live', (Royal Commission into Misconduct in Banking, Superannuation and Financial Services Industry, Interim Report, Volume 2: Case Studies, p. 361) and to keep their machinery to allow them to make a living. On 31 January 2012, internal bank documents noted that should the houses be excluded from the auction, the sale price was unlikely to be affected. On 1 February, the auction proceeded and one Cheesman home was sold. A debt of $630,000 remained. The family sold machinery and equipment in a desperate effort to keep one house for both families. It was a stressful battle which they finally won, but their livelihood was lost.

The Department of Agriculture and Water Resources announced that bank lending accounted for ninety-five per cent of total institutional lending to the agricultural, forestry and fishing industries and more than ninety-five per cent of broadacre and dairy farms were family owned and operated. The total indebtedness as of June 2016 was $69.5 billion. (www.farmonline.com.au/story/4788225)

Janine and Stephen Harley, of Western Australia, were also Landmark, then ANZ, clients. Their family had been custodians of their land for a hundred years. Despite tough weather-related years, they believed they were months away from clearing their debt. One of their properties was sold to pay down the bank debt in July 2012. The stress began to damage Stephen Harley's health. In October 2012, he was hospitalised with pneumonia. By January 2013, the sale of the entire holding failed to find a buyer, so in March the family subdivided and offered all for sale. The properties did not sell, so the debt could not be settled by the March 2013 bank deadline. In May 2013, Stephen suffered a heart attack and Janine Harley notified bank authorities. The bank agreed to an extension with the proviso that unless the debt was met the Harleys would give the ANZ vacant possession by 1 April 2014.

By 27 March 2014, the family had sold five of their nine parcels of land as well as livestock. They had paid off $1.6 million of the $2.5 million debt and asked for another extension so remaining properties, other than the one on which their home stood, could be sold. They argued that it was better financially to sell the properties in spring rather than winter and that their continued care could also secure a better price. This was refused and the ANZ appointed agents for a mortgagee. The Harleys were evicted and told they were required to pay for the mortgagee agents. The fire sale proceeded immediately and when the debt was not attained from the sale, the family was threatened with bankruptcy proceedings. The pressure to resolve their financial issues had been immense. The Royal Commission was told by bank officials that 'ANZ did not accept that its conduct…fell short of community standards and expectations'. (Interim Report, Vol. 2, p. 371)

A secret ANZ document revealed that lending services staff were encouraged to remove distressed farms from the bank. A key performance indicator (KPI) was to finish up files within two years. Banking staff bonuses increased if this was achieved. ANZ admitted it took enforcement action within four years against thirty farm businesses. The Commonwealth Bank (CBA) admitted that it had taken enforcement action against eighty-two agricultural customers during the past decade. It appeared that responsibility for poor financial advice given to farming families was never admitted and covered up with a rapid foreclosure, fire sale and eviction.

In 2009, Queensland cattle farmers Wendy and Adrian Brauer were actually alerted by their Rabobank manager, that another property, 'Jamberoo', eighty kilometres from their own, 'Kia Ora', was on the market and offered a $3 million loan for the purchase and an additional $300,000 for more stock. They were assured of their ability to service the loans and proceeded with the acquisition. The Rabobank manager had miscalculated. The properties suffered flooding and the Australian government's ban on the live export of cattle to Indonesia in 2011 deepened the debt. Wendy Brauer 'enquired as to hardship arrangements' with the bank manager and was told 'no such arrangements were available'. (Interim Report, Vol. 2, p. 396) Default interest of four per cent above the standard rate was applied and the debt deepened further. The Brauers engaged Legal Aid, and Rabobank documentation was requested and questions asked concerning bank employee conduct. Rabobank declined to provide documents and refused to be questioned. The Brauers were forced to sell the property at a loss to them of $1 million.

Rabobank later admitted that rural managers were 'incentivised' to grow new business loans and 'disincentivised' to write unsustainable loans. Two uncommon words had taken priority in the banking business. For the banking community, it meant, 'to make someone want to do something, such as to buy something, or to do work, especially by offering prizes or rewards'. For the unsuspecting public, it meant the friendly bank manager with your best interests in mind had long gone – it was now all about business greed and personal profit.

Among the final 10,140 submissions received by the Royal Commission, 288 related to agriculture and rural debt. Commission Hayne extended the time allowed for farm case studies, but the Australian government expected an interim report around September 2018 and the final report was due on 1 February 2019. The legal requirements and

Commissioner Hayne

arguments took time, a great deal of time. Financial authorities being interrogated answered with careful and time-consuming consideration. Bill Mott was concerned that there was not enough time for him and other farmers to speak. 'It is all about the outcome, trying to get rid of us, it is never about our experience.'

In Brisbane, however, he had been made aware of others and was welcomed into the band now calling themselves the Bank Warriors. There was strength in numbers and company on this journey was welcome. This group of Australians were to spend another year gathering in Canberra to make their presence known in parliamentary committee hearings – determined to support each other through black days and days of minor victory.

Bill Mott was resolved to prevail; he had every reason to. In Brisbane, he moved from the Royal Commission hearings to another courthouse to finalise his divorce.

> Not only has our family lost millions of dollars but we have lost our home and our lifestyle, my children have lost their future inheritance and had to choose new careers, I have been divorced and we have all endured significant depression and embarrassment.

The journey was longer and more hard-fought than he could have imagined.

Nine

'It should not be only about money, but about people and the future of our kids.'

Bill Mott had never travelled far from Meandarra and now he was becoming very familiar with capital cities. He started with the NAB head office in Brisbane. His detailed letters received courteous response, but nothing more. The plea 'I do believe time is of the essence' went unheeded. His large four-wheel drive vehicle with antennas and giant bull bar was intimidating in the city traffic, and the farmer manoeuvring it was beyond being bullied. He couldn't get the head office NAB chiefs to take his numerous phone calls, so he drove to Melbourne and hand delivered a letter to the chairman, Ken Henry.

It was difficult and complicated to navigate the morass and at every turn there seemed another obstacle to understand and overcome. A Hong Kong insolvency lawyer was brought in. His charge was $100 a minute; 'he played with me for three months, couldn't prove anything against me so turned the case back to NAB.' It only strengthened the question as to why the bank was so intent on discrediting Bill, at great expense, avoiding supplying documents requested and delaying disclosure concerning its dealing with the Mott family. He was granted a meeting with a high profile bank official, a couple of meetings, but 'I was spinning my wheels…he continued to say nothing had been done wrong.' Then there was the realisation that it was 'not what he said but what he didn't say'.

For a man who never sought more than working broad Queensland acres, a close-knit family and a modest lifestyle, this was all new and out of character. In the beginning, it was supremely difficult to enter an unwelcoming hearing room and, surrounded by officials in black suits and with solemn expressions, eloquently engage their interest if not concern. The passion was there, the anger barely masked, and in his own words, he was 'a very determined person' and he was sure he had been a victim of fraud. For a farmer of few words, he appreciated he needed to be heard by the media, not for himself alone but for his family and the other farmers and their families. It took time to be comfortable with the scrutiny – if it ever was comfortable.

He was grateful to be part of the Bank Warriors, a band of people with one thing in common: they had been brutalised by financial institutions. They were from different corners of the nation, different banks, different farms, but had been subjected to similar treatment. They gathered in distant cities in the same modestly priced motels, talked over the odd beer and counter meal and, most importantly, were there for each other, in that odd awkward silent way of men. They had lost everything, so what more did they have to lose.

One farmer spoke of his wife's anxiety whenever he took out a rifle to shoot a fox or an injured sheep, as she prayed their financial situation did not result in his using the rifle on himself. Another farmer alleged a banking representative suggested that if he found himself 'in my shoes they would commit suicide'. (*The New Daily*, 1 December 2017) They acknowledged shared responsibility – they had signed the loan documents – but standard bank practice could never fit agriculture. The recurring themes were too much credit; manipulated loans; default due to factors beyond a farmer's control; and the uncompromising and callous treatment of banks when they crushed farmers with default interest of three to four per cent on top of payments they couldn't pay already.

The report which had been submitted by the Senate's Select Committee on Lending to Primary Production Customers in October 2017 suggested that it was impossible for agriculture to be treated in the same way as small business because, on average, it took one in three years for a farmer to break even or make a profit and reduce debt. The report had made twenty-seven recommendations, mostly directed at the Australian Bankers Association (ABA). The ABA had even supported a mandatory National Farm Debt Mediation scheme; that 'fairer and more transparent practices' be introduced so that farmers were 'treated fairly'. (*The New Daily*, 1 December 2017)

Why had not something been learnt from the stories brought to the surface, and recommendations made, by the 2017 Senate inquiry? Three members of the committee – chair Senator Malcolm Roberts, deputy chair John Williams and Senator Claire Moore – could painfully recall the agonising testimony in the Commercial Hotel in the north Queensland inland town of Charters Towers in July 2017. Charters Towers epitomised the changed face of country Australia, and stories of boom and bust. The boom had occurred between 1872 and 1899 in the form of so much gold that the town boasted its own stock exchange and a population of around 30,000. By 2016, the population was around 8,000 and mostly dependent on agriculture.

Dennis Fahey had described to the committee how he had catered for market concerns by establishing an organic cattle business. It was valued at around half a billion dollars when the bank foreclosed in 2011. His marriage broke down, his son was suicidal, and he had been forced into hiding.

> The banks don't care. This behaviour is fraud and is criminal and it has knocked us around something shocking. The divorces, the family breakups, the depression, the suicides, your guts hits the floor. (*Townsville Bulletin*, 14 July 2017)

Dennis Fahey now survived on a pension.

If physical proof was needed, it was blatantly obvious in the burn scars carried over fifty per cent of Brett Fallon's body. He had owned cane and cattle properties, and then he had nothing. In April 2013, Fallon was destitute and desperate, with no control over his life. He had gone through nine months of grief, 'trying to get answers from accountants …trying to track down funds I knew was owed to me'. Doors were shut and contact broken off – there seemed nowhere else to go. He had a smouldering fire at the back of his property. 'I tipped 1.5 litres of petrol on me and walked into the fire.' He survived after seven months in a coma. 'People have been dragged off their properties at gunpoint… These banks are deliberately choking these farmers to death financially.' (*Townsville Bulletin*, 14 July 2017) By January 2018, the fifty-eight-year-old Fallon lived outside Bowen, Queensland, in a hut which was once a sale yards cafeteria. His former partner Julie, wrote,

> I also lost everything…including children. For that loss a cheque simply doesn't go near compensation. Only death ends that loss. (www.bankvictims.com.au)

It became clear that the Royal Commission into Banking Misconduct had been hobbled by its terms of reference. It was not allowed to include these tragic stories nor anything it was believed had been, was being, or would be, appropriately and sufficiently and dealt with by any other investigation or inquiry, or criminal or civil proceeding. So the 2017 Senate report of and the horrors included, were to be ignored and so too, the findings of the thirty-eight or more previous inquiries held into banking and financial services since 2010.

To ignore what had already been disclosed meant the commission could not truly comprehend the one-way nature of bank contracts and a system whereby borrowers remained indebted to their banks even after foreclosure. In the United States, non-recourse lending meant borrowers whose properties were forcibly sold off had no ongoing bank debt. This also encouraged greater caution amongst lenders. The commission's terms of reference also did not encompass the lack of transparency of insolvency professionals involved in farm foreclosures, nor the need for a national farm debt mediation scheme where banks were forced to mediate with farmers.

There needed to be discussion about the establishment of a rural reconstruction board, which could act as guarantor for struggling agricultural businesses, securing low borrowing rates and setting reasonable repayment conditions. So many issues with direct bearing on rural communities, yet none could be laid bare. National Farmers Federation president, Fiona Simson, declared the time allocated was unacceptable; much more was needed to even attempt to unravel the complicated scenarios faced by farmers.

During the June commission hearings in Brisbane, counsel assisting, Rowena Orr QC, announced that the focus needed to be on the 6.9 million Australians living in regional and remote communities as of 30 June 2017. This meant that around twenty-eight per cent of the population interacted differently with financial institutions than those in metropolitan areas because the closure of branches, lack of internet access and higher ATM fees. Orr also agreed that farmers faced very different financial challenges.

Mel Ruddy was invited to state his case at the commission hearing. Ruddy was a cattle breeder near Charleville in western Queensland. With a sizeable loan and with cattle prices falling in 2011, Ruddy had decided to sell one cattle property to focus on another. A Bank West branch manager who boasted of now having his valuer's ticket revalued the Ruddy properties and convinced the family to take out a 'better loan' with the bank rather than sell a property. They were assured that the properties were worth $2.3 million, and the bank was happy to lend sixty per cent of that value with extra equity in tough years. Low rain and the live cattle export ban tightened 'the fiscal screws' and 'we were struggling'. (*Financial Review*, 28 June 2018) Cattle were sold at much reduced prices.

The friendly bank manager responsible for the Ruddy family

dilemma moved away and the loans were then handled by Commonwealth Bank employees 800 kilometres away in Brisbane. The Ruddy properties were revalued at $1.65 million. To dig themselves out of debt, the family sold their primary property and home, and equipment, and borrowed $160,000 off Mel Ruddy's mother. Under cross-examination by Orr, a Commonwealth Bank executive admitted that the bank manager concerned had resigned in 2013 when it was found he had overstated valuations and had been involved in 'other discrepancies'. (*Financial Review*, 28 June 2018)

The Australian Banking Association (ABA) declared the nation's agricultural industry was 'strong and has a positive long-term outlook' and 'There is no financial incentive for a bank to deliberately undervalue an asset or lose a customer.' (www.farmonline.com.au/story/4788225) National Senator John Williams called for tough penalties and criminal proceedings against bank employees and liquidators. (*The Guardian*, 23 June 2018) The government had promised since 2016 to introduce legislation to allow the Australian Securities and Investments Commission (ASIC) powers to intercede in dubious financial practices but by 2019 had failed to do so.

As the Royal Commission picked up pace, the Commonwealth Bank chief executive, Matt Comyn, took the opportunity to visit country areas to hear directly from the rural community, and had promised the bank would improve its relations with farmers. But Comyn admitted he had met with fewer than ten victims of bank misconduct. He also stated that 'there were some 45,000 employees in the Commonwealth Bank, and you can't look over the shoulder of each and every one of them'. (*Investor daily*, 11 October 2018) Unfortunately for Comyn and other bank chief executives, during the ensuing months it would be disclosed just how persuasive the culture was and how much emanated from those in charge.

In August 2018, around 150 people crowded into the main committee room of Parliament House, determined to make their presence felt in support of the victims of financial abuse offering testimony. They also had come to make known their discontent, with the inadequacies of the commission. Bill Mott was not alone in his disappointment:

> I initially put a lot of faith in the Royal Commission, but now I see it's really not going to provide compensation, it's not going to fix problems – it may stop problems in the future but it's certainly not going to fix the ones out there now.

He was with the cross-section of Australians who moved to the Senate public gallery for the debate. Yet political support for changes to the inquiry did not echo from the two major parties in the chamber below, but from politicians who were commonly labelled 'mavericks', even 'right-wing mavericks' – the same politicians who had galvanised support for the Royal Commission the year before.

National MP George Christensen believed, 'It would be a tragic lost opportunity if the full extent of deliberately dodgy banking practices were not exposed.' (*The Guardian*, 23 June 2018) By June, there had already been 6,761 public submissions to the commission yet just sixteen members of the public had been invited to speak. Each bank called was asked to outline any misconduct or conduct falling below community standards, committed during the last decade, in a document not exceeding fifty pages.

In the Senate chamber, Bob Katter's Australia Party Senator Fraser Anning declared the Royal Commission was underfunded, did not have adequate time to hear submissions, and the terms of reference needed to include all financial professionals who acted unconscionably and

> Extend the final reporting period by 12 months to enable the commission to hear more submissions and increase funding to the royal commission to enable it to hear more submissions…it needs not to be a white-wash…we don't want a quickie royal commission. (*ABC News*, 2 August 2018)

National Party Senator John 'Wacka' Williams had long campaigned for the victims of banks, defying his coalition partners in calling for a Royal Commission.

The whole process has not been long enough. My attitude is do it once, do it properly. Nothing would be worse than to see that it was rushed, and they could have done a better job. (*Farm Weekly*, 8 August 2018)

Senator Pauline Hanson (Pauline Hanson's One Nation Party) declared that all Australians needed to hear victim stories so that the community could decide on what action should be taken against banking executives. Commission hearings needed to be expanded so 'no Australian ever has to suffer through this mistreatment at the hands of the banks'. (Parliament of Australia, Hansard, Senate, 27 June 2018) She also believed that the cost of the commission should be passed on to the financial institutions found guilty of misconduct.

Labor Party Senator Chris Ketter believed,

There was a lack of consultation on those terms of reference, which would've been better handled if the victims had been consulted. In fact, it would've been, probably, advisable for ASIC to have been informed and to have had an input into those terms of reference. But that didn't happen, because we know that the banks came out one morning at about 8.30, in writing, to indicate that they now supported a royal commission, and it was something like an hour later that the Prime Minister did one of the most enormous backflips in political history. (Parliament of Australia, Hansard, Senate, 27 June 2018)

Labor Senator Malarndirri McCarthy supported her colleague and agreed that the Turnbull government never wanted the Royal Commission. She was concerned that this was 'to protect their mates in big business…to help their mates working for the big banks'. (Parliament of Australia, Hansard, Senate, 27 June 2018) Senator Peter Georgiou (Pauline Hanson's One Nation Party) concurred; he believed the banks and the government had indeed colluded in the writing of the terms of reference. National Party Senator Barry O'Sullivan added, 'The banks are on notice…we need the Government to support this inquiry if it needs more time and more resources to do a thorough job.' (Parliament of Australia, Hansard, Senate, 27 June 2018)

Assistant minister to the prime minister, Senator James McGrath, defended the government, declaring that the commission was independent of government and could decide to examine what it wished. The government was happy with the commission's progress, he said. The terms of reference nonetheless had been set by the government, as had the requirement for an interim report in September and the final report by February 2019. Regardless of the support within the Senate, neither the terms nor the time were extended. Bill Mott was unimpressed with the progress.

> They're not going to scrape the surface. It's very disappointing to think that they've had that number of submissions and that's the best time they can offer, because there's real pain.

He was becoming more confident in these alien surroundings and less inhibited by officials. There were sympathetic interviews with two influential opposition politicians – Clare O'Neill MP, Labor's financial services spokeswoman and a member of the Agriculture and Industry and Tax and Revenue Committee; and Matt Keogh MP, a former commercial litigation lawyer and a member of the House Economics Committee, the House Agriculture & Water Committee and the Committee on Corporations and Financial Services.

A meeting with the Government Financial Services minister, Kelly O'Dwyer, was not as productive. O'Dwyer simply espoused the party line: the commission was not only independent but was doing a good job. Her response, that if farmers felt they had been treated unfairly they should hire legal representation, was out of touch with reality.

There was real relief when Bill linked up with Lynton Freeman. Freeman had fought his own battle with NAB through the courts. His working experience included the Queensland Justice Department, local government, agricultural scheme administration and twenty-five years as a farmer, grazier and silviculturist. In addition, his academic qualifications in agriculture and rural business management, business administration and global law, made him a very knowledgeable and formidable associate. Bill had struggled with the complex financial data. Lynton skillfully undertook a forensic accounting analysis of the Mott/NAB transactions and both men were amazed by the level of fraud discovered.

NAB's transaction with the Mott family had, among other irregularities, involved bank bills. The bank bill swap rate (BBSW) is used to settle bank debts with other banks and how financial products are priced. The NAB could borrow from an overseas bank at a low rate but charged Australian customers a much higher rate. The Motts' interest rate was based on this BBSW but then increased with extra interest. Some of the BBSW was to be paid to the Australian Tax Office but financial accounting could be delayed and altered. In 2017, NAB was required to pay a $50 million penalty by ASIC after an investigation discovered that NAB staff were found to have broken the law by engineering the interest rate between 2010 and 2011. The same fine was administered to the ANZ, whereas the CBA was fined $25 million for bank bill manipulation. Westpac chose to fight the penalty in court. ASIC proved that Westpac had attempted to manipulate the bank bill rate a number of times in 2010 but had not succeeded. Westpac was fined $3.3 million for 'unconscionable conduct and breaches of the obligations of its financial services licence'. (Courtesy of Daniel Ziffer, ABC)

It had been a long year and so many farmers and small business owners were simply worn down too much to continue. These were proud, hard-working Australians who had been taken advantage of by lesser individuals. Bill couldn't give up – he had nothing left to live for but this fight. It had never purely been about the money it was about much more. 'It should not be only about money, but about people and the future of our kids.'

Ten

'They railroaded me to lose my powers and I only got power back through the Standing Committee inquiry.'

There were high hopes for the Royal Commission into Banking Misconduct, and the hearings were followed closely by victims. Commissioner Hayne and Rowena Orr asked hard questions of financial officials, whose expressions and comments revealed their discomfort. As the shocking stories of misconduct, mismanagement and greed were revealed, few remained unscathed. Some of the worst cases came from the commission's scrutiny of NAB's introducer program. NAB bankers were said to have accepted loan bribes, were given envelopes 'stuffed with cash', forged customers' signatures and manipulated incentive programs to generate bonus payments. NAB emerged a key target of the commission interim report released on 28 September 2018. Hayne pilloried NAB for the inability of board members to appreciate their institution's wrongdoing, and for showing disrespect for the gravity of the findings.

A key finding was that NAB had too often focused on what was 'expedient' for the bank rather than what was best for the customer. NAB chief executive Andrew Thorburn admitted the failures of the bank and promised such mistakes would not occur again. (*The Canberra Times*, 27 November 2018) His bank had 'too often failed to put clients first'. (*The Canberra Times*, 1 December 2018) The scolded NAB chairman, Ken Henry was forced to admit, 'too often the voice of the customer has been missing from our decision-making'. (*The Canberra Times*, 1 December 2018) He described the behaviour as the fault of capitalism. 'It goes to the state of capitalism. The capitalist model is that businesses have no responsibility other than to maximise profits to shareholders.' (*The Canberra Times*, 28 November 2018) Customers were a means to an end.

The interim report only confirmed what Bill Mott and other Bank Warriors had known. They now looked to the ongoing Australian Parliamentary Standing Committee on Economics: Review of the Four Major Banks, to reveal better the human face of the banking crisis, to ask the uncompromised questions of the nation's top banking officials and for perhaps their own opportunity to corner the CEOs. This committee was a bipartisan one with wider terms of reference and with the power to call whoever members wished. The committee's third report had been delivered and the fourth and final report was due in December 2018. No bank escaped committee scrutiny. Westpac had been the most resistant of the four major banks to the supervision of ASIC, ignoring requests to compensate customers who were inappropriately sold loans. The ANZ came under fire for weak loan controls. It was cited that NAB bankers accepted loan bribes, forged customers' signatures and manipulated incentive programs to generate bonus payments. NAB had also been slow to compensate victims.

Labor MP Clare O'Neil was aware of the Mott family saga and assured Bill that difficult questions would be asked. The CBA's Matt Comyn was the first CEO to face the committee. O'Neill wanted the CEO to understand.

> You are probably aware that of 10,000 submissions, just 27 people have been allowed to tell their stories [at the Royal Commission]. I don't think that is sufficient and as a consequence, I am travelling around Australia with the help of some of my colleagues to visit people who have been victims of bank misconduct. (*My Business*, 11 October 2018)

The MP expressed her concern for people who had been 'incredibly hurt' by the actions of Australian banks.

> I have met a lot of people who have had very significant mental health issues…from the treatment by your bank and others. I have spoken to people who have anxiety and stress, people who have attempted suicide…farmers who have lost farms that have been in their families for generations. (*Investor daily*, 11 October 2018)

Comyn could but concede that there had been 'failures of judgment, failures of process, failures of leadership and in some instances, greed'.

Bill Mott arrived at Canberra's Parliament House early on Friday 19 October 2018 because this was finally the day NAB CEO Andrew Thorburn and David Gall, the NAB chief customer officer of corporate and institutional banking, were to appear. Bill did not blend easily in these surroundings. His long-sleeved shirt and fawn chinos were neatly pressed and the boots, more at home treading the black and red dirt of the Western Downs, were polished. The jacket was discarded early, and the tie looked uncomfortable. The Australian parliamentary building is an alien place – it is intended to represent the nation's people, but it fails to make them feel welcome. The towering foyer with its marble columns and staircases is formidable. Bill and other Bank Warriors had been there so many times the security guards were familiar acquaintances. However, these Australians with few assets to their names continued to feel ill at ease as they walked down the long corridor festooned with stern portraits of past prime ministers to the main hearing room.

The main hearing room is impressive. A massive triptych fills the high front wall behind the committee seats. The contemporary interpretation reflects the strong colours of an Australian seascape. Eyes are attracted to the thick stone pillars that dominate the painting, ironically, they could represent the banks.

The main committee hearing room, Parliament House

The media with long camera lenses positioned themselves in front of the CEO. The committee chairman warned the murmuring audience settling into their seats that they were not to interrupt proceedings in any way – only committee members could ask questions.

In his opening statement, Thorburn admitted the Royal Commission had exposed issues within the NAB that he had found 'confronting and upsetting', but he went on to highlight the special concessions NAB had made to agribusiness customers, including ending the use of default interest rates for farmers in drought-declared areas. When the committee pressed the CEO on why this did not seem to be the case for all drought-affected areas, the answer lacked clarity and the CEO admitted he was awaiting a broader review of default interest rates, and that the bank's business had 'become too complex'. (www.aph.gov.au/Parliamentary_Business/Committees/House/Economics/FourMajorBanksReview4)

It was pointed out to the NAB bankers that the Royal Commission Interim Report noted that NAB had reviewed 11,000 files and had identified about 1,360 customers who might have been affected by the misconduct. Yet NAB management believed the amount it expected to pay out was between $9 million and $23 million. It was difficult for Bill Mott to maintain his silence – his family alone were owed most of that $23 million. On behalf of the Mott family, a committee member asked why the NAB had 'charged a default interest rate of eighteen per cent to a farming family as they struggled to recover from flooding'. The specific case was unfamiliar to the CEO.

Thorburn stated that he had been in banking for thirty years and admitted that over the last twenty years the banking business had changed.

> It looks incrementally sensible; others are doing it and there's a system that reinforces it. Shareholders appreciate it if you're making good and better profits. (www.aph.gov.au/Parliamentary_Business/Committees/House/Economics/FourMajorBanksReview4)

He assured the committee that he would implement cultural change and that 700 NAB staff members had been given a pay reduction and more than 300 had resigned or been terminated as a result of investigations into their conduct – just one per cent of NAB staff. However, he was certain that the majority of the NAB's 33,000 staff acted with integrity. The usual characterisation of 'a few bad apples' did nothing to convince the audience, because it not only attempted to diminish the extent of the misconduct, but it also distanced the entity from responsibility and the NAB's systemic failures in processes to manage risk and compliance.

On 19 October 2018, NAB announced an additional $314 million for refunds and compensation to customers. (*ABC News*, 19 October 2019) It was to prove not nearly enough.

Thorburn ended his testimony with an offer:

> If there are National Australia Bank customers, I would be happy for them to contact me directly… What they should expect from me and from our company is a prompt response and a review within a reasonable period of time. (www.aph.gov.au/Parliamentary_Business/Committees/House/Economics/FourMajorBanks Review4)

It was the offer and opportunity Bill Mott had waited years and travelled thousands of kilometres for. 'They had railroaded me to lose my powers and I only got power back through the Standing Committee inquiry.' He immediately approached both Thorburn and Gall with Lynton Freeman and the files Lynton had undertaken a month-long forensic analysis of. They were confident this demonstrated how NAB had manipulated bank bills and tampered with interest. Thorburn's signature was on three of the twenty-eight pages. The meeting was to be fifteen minutes but lasted twice that. Thorburn showed emotion as Bill relayed his sad story. 'I don't think he realised the extent of the damage that had been caused.' Bill believed it demonstrated that bankers like Thorburn 'have got a heart'. NAB chief customer officer of corporate and institutional banking, David Gaul, admitted to Bill Mott, 'we know we owe you money, but you are asking too much'. He assured Bill that the NAB documentation previously denied would be forthcoming in the immediate future.

Thorburn shook his hand and as Bill stepped out of that meeting he felt more positive than he had for three years. 'Andrew Thorburn had a great attitude…just a total turnaround from the other meetings where all they wanted to do was lie and cover stuff up.' He really believed Thorburn had listened, assured him the matter would be resolved, and that there would be contact 'within a fortnight'. 'It was a good result.'

Bill climbed into his large four-wheel drive and the trip north to Queensland was far more enjoyable than it had ever been. It was difficult to contain the optimism. A top official had admitted the Mott family was owed money. Finally, there was vindication. For so long he had believed he had failed his family, but the NAB chief assured him he would personally oversee the financial resolution Bill had fought so doggedly for and there would be contact 'within a fortnight'.

Eleven

'Andrew Thorburn had a great attitude…just a total turnaround from the other meetings where all they wanted to do was lie and cover stuff up.'

Sitting at the bar in Meandarra's only pub, Bill Mott was having a steak sandwich. Through the window, the concrete war memorial is clearly in view. Beside it a Lone Pine sapling struggles to survive too many years of drought in a climate and soil even harsher than from where the parent tree overlooked the grim battles and bloodshed on a Turkish peninsula in 1915.

Like the tiny sapling outside, life had been a struggle for this man of the land. The sandwich wasn't entirely satisfying nor the shandy which stood on the towel-covered bar. He had been promised a phone call by NAB chief executive Andrew Thorburn 'within a fortnight' and a fortnight had been and gone. The stress in his chest was rising. Bill at least could shelter from the heat outside, the intense heat of Queensland's Western Downs in November, unrelenting dry heat, which would only intensify in the ensuing months to stubbornly linger in the forty-plus Celsius range. Farmers and graziers had ceased looking up, hopeful of finding rain-bearing clouds.

The hotel was one of very few places where people could mingle, but it too came close to shutting its doors. A family bought it at the eleventh hour, much to the relief of the locals. Opposite there is a large business with signs that signify what is happening in this town, most country towns – 'Closing Down, Everything Must Go'. The conversation in the bar revolves around the weather, crops and cattle and equipment. There is a mixture of bluster, old jokes and despair. For visitors, there was bluster and dry Australian humour. 'You think this is hot, just wait' and then something about heat that melts faces off. There is the shearing story about when the wool industry prospered, and shearers were lining up in great numbers at 'Bindebango'. They camped out on the clay pan until someone quit and the next in line took their place. A couple got word they were to start so they turned up at the property kitchen for breakfast. A station hand had killed a cow and the chops served came from that animal. When the two men saw the size of the chop on their plate, they got up and left muttering, 'If the chops are that big, we're not shearing these bloody sheep!'

Bill relates his own tale. After three years at 'Homeboin', the road into town was increasingly rough because the council had not graded it for that length of time. The next time he paid his rates bill, he wrote on the 'list complaints and requests section', 'Could you please send a photo of a "grader at work" as I cannot remember what one looks like.' The road was graded the following week. Members of the local council used to be elected volunteers. 'Then they got a big salary and car and lost touch with and interest in the local community.'

A couple of gas workers distinguished by their high-vis shirts sit down for lunch. There is an uneasy truce between them and the men and women of the land. Mining has monopolised the railways, making it impossible to transport agricultural produce. The presence of gas workers assists local services but farming families have concerns. Fracking is the main one. Fracking is the process to release and increase the flow of natural gas trapped underground. Under high pressure, fluid is pumped down a well to open cracks in the rock. The fluid contains chemicals and sand and once it re-emerges on the surface, it is supposedly held safely in deep ponds. The chemicals are toxic. The concern is that in flooding rain the ponds can overflow into waterways and enter the water table. Bill Mott had summer crops ruined by twenty-six inches of rain in almost as many hours. His wedding was flooded out and guests were marooned by overflowing waterways. The concern is very real.

The steak in the sandwich becomes an issue of conjecture. Australia's best meat is all shipped overseas, it is disclosed. There is a rule, 'if it has a bone in it, then it is likely Aussie beef because meat with bone is not allowed to be imported'. The inevitable question: 'Why does Australia export the good stuff and import the inferior stuff?' The mood turns gloomy. A common lament is that there is no longer the pride in agriculture that families took. Planting is rushed, done by contractors as cheaply and quickly as they can for distant owners, resulting in poor quality farming. This country used to be generational family properties, 'What happened? Why is it increasingly owned by foreign governments and corporations?'

There are allegations aplenty here about foreign ownership, agricultural mismanagement and the run-on effects for Australian rural communities. The United Arab Emirates (UAE) purchased a large holding for cattle. It was believed the Australian way was unproductive, so the fences were ripped up and temporary fencing and drovers utilised, like they did back in the UAE. The method proved very unsuccessful, with animals straying. That foreign consortium moved away. 'Funny that' how since even the first colonists invaded the land now known as Australia, no one took advice from the locals on what was good for the land. All the stuff introduced which later proved disastrous for the nation's ecology, from rabbits and foxes to cane toads to prickly pear – the list goes on and on.

The Australian rural sector is under attack from everything but, after the banks and mining, foreign leasehold and ownership of Australian agricultural land is foremost, because it could in decades endanger the country's ability to feed its people. When Bill Mott's grandfather, William Mott, accepted a soldier settlement in far western Queensland, his nation's population numbered less than seven and a half million. In 2017, it was more than twenty-four and a half million. Yet Australians own much less of their agriculturally productive land and have less food and water security than in 1945.

In December 2010, it was estimated that 11.3 per cent or 44.9 million hectares out of 398 million hectares was foreign-owned. By June 2013, 12.4 per cent or 49.5 million was foreign owned. (*ABC News*, 7 September 2016) On 7 September 2016, then-Treasurer Scott Morrison released Australia's first report from the Agricultural Land Register, 'to increase scrutiny and transparency in Australian agriculture'. He stated,

> Foreign investment is integral to Australia's economy. It contributes to growth, productivity and creates jobs, but the community must have confidence that this investment is in the national interest. (http://sjm.ministers.treasury.gov.au/media-release/092-2016/)

In 2016, 13.6 per cent of Australia's agricultural land was held by international investors. More than fifty-two per cent was owned by United Kingdom investors. Countries with the next largest shares were the United States, Netherlands, Singapore and China. It was accentuated in the media release that 'less than half a per cent (0.38 per cent) of Australia's agricultural land is held by Chinese companies'. (http://sjm.ministers.treasury.gov.au/media-release/092-2016/) In 2016, of the foreign-controlled land, 9.4 million hectares was freehold, 43.4 million hectares was leasehold. Livestock followed by cropping were the most popular agricultural pursuits, while the highest portion of foreign-owned agricultural land was in Queensland, followed by the Northern Territory, Western Australia and then South Australia.

It was 7 September 2016 when Treasurer Scott Morrison released the first report from the Agricultural Land Register, 'to increase scrutiny and transparency in Australian agriculture' and September the following year when the next report was released. The third report was released not in September but on 20 December 2018. By 20 December 2018, the Australian parliament had risen weeks before and were expected to sit just thirteen days before a May federal election. The federal government had commenced the summer shutdown. Schools disgorged the final students for the six-week vacation and families had taken to Australian highways and air to be with relatives and friends for Christmas and New Year.

Few Australians were settled enough or had time in the pre-Christmas last-minute buying frenzy to read the media reports or comprehend what was happening. 'Chinese buy more Aussie acreage… Chinese investors have added another 50,000 hectares to their Australian property portfolio', reported one newspaper. (*The Canberra Times*, 21 December 2018) *The West Australian* described how foreign ownership of agricultural land had grown at the highest rate in the western state. It was now about 6.9 per cent or 890,000 hectares. (*The West Australian*, 21 December 2018) This meant that 13.7 million hectares of Western Australian farmland had a level of foreign ownership – 16.4 per cent of the state's total agricultural land. Of this, 7.9 per cent was freehold and 92.1 was leasehold.

The UK remained the largest foreign agricultural concern in Australia adding 487,000 hectares during 2017 for a total of 10.2 million hectares. (*The Weekly Times*, 21 December 2018) Whereas China, which held control over 1.5 million hectares in 2016, now controlled 9.17 million hectares.

Strangely, the Bahamas, or at least a company or individual registered in the tax haven, had undertaken 'the mysterious purchase of more than 2 million Australian hectares, almost twice the size of the Caribbean island itself'. (*The Canberra Times*, 21 December 2018) The Australian treasurer, Josh Frydenberg, repeated the same statement made by his predecessor, Scott Morrison in 2016: 'Foreign investment was an important contributor to growth, productivity and jobs in agricultural communities.' This could not be further than the truth, at least according to rural communities in western Queensland.

There had been hopes that when a large Chinese consortium established itself on 120,000 acres of Western Downs country, it would bolster the local economy and ease unemployment. This did not occur because workers and everything to sustain them were brought in from Asia. The workers were housed in specially built onsite accommodation. The roadhouse the Meppem family had called home was then purchased to exclusively cater for them. A purchased abattoir allowed for the processing and packaging in-house and the produce was then delivered by their own trucks to be loaded on ships owned by the same interest and transported to the home country. Locals wondered how any of this 'was an important contributor to growth, productivity and jobs in agricultural communities'?

Over the previous decade, family holdings had been lost. In 2009, local and overseas corporations, led by Macquarie Group and the British takeover specialist Terra Firma Capital Partners, engaged in a frenzy in the Australian north which resulted in the most hectic period of property activity in the 185-year history of grazing. Contracts for more than $1.1 billion were exchanged on more than thirty of the north's best-known cattle stations. (*Brisbane Times*, 21 March 2009) In 2016, the 117-year-old family empire founded by Sir Sidney Kidman, with pastoral leases covering 101,000 square kilometres across Western Australia, Queensland, the Northern Territory and South Australia – 1.3 per cent of Australia's land mass – was placed on the market. A bidding war ensued between Chinese Genius Link Group and Chinese company Shanghai Pengxin. The then-treasurer blocked the sale but agreed to the cattle empire being sold to a joint venture between Gina Rinehart's Hancock Prospecting and the Chinese Shanghai CRED for $385.6 million. The overseas interest was not universally popular. MP Bob Katter felt 'absolute rage and disgust' that the Turnbull government was 'selling their country out and selling their country off'. (*ABC News*,

10 December 2016) WA Farmers chief executive Trevor Whittington noticed the emerging trend.

Western Australia has been experiencing consolidation across our grain belt for decades as small 1000 acre properties become 2000, 3000 and upwards as people leave the land. (*WA today*, 21 December 2018)

In August 2018, it was reported that Shimao Property Holdings, controlled by Chinese-Australian billionaire Hui Wing Mao, which operated sixteen cattle stations valued at $1 billion, was one of 'several high-powered international parties including Chinese bidders circling the sprawling Consolidated Pastoral cattle empire (CPC)'. (*The Sydney Morning Herald*, 18 August 2018) It was believed the Australian Government blocked the outright sale to foreign interests. However, the cattle empire was purchased by 'an Australian managed, majority foreign owned agrifood business'. (*Farm* online, 9 January 2019 Three stations were then sold but CPC's portfolio included twelve cattle stations with a carrying capacity of over 325,000 head across 3.9 million hectares of Australia. It also held two feedlots in Indonesia. The beef from the largest holdings in Australia, such as Consolidated Pastoral and Kidman, is exported.

In 2013, the massive Queensland Cubbie Station, the southern hemisphere's largest irrigation farm, and St George properties ('Anchorage' & 'Aspen') were sold to a Chinese-led consortium, textile giant Shandong Ruyi, taking an eighty per cent share, and a Melbourne-based family company, Lempriere, holding the remaining twenty per cent interest. It encompassed 80,190 hectares, an amalgam of twelve former sheep and cattle properties. The Dirranbandi cotton gin was purchased by the consortium in 2013. According to its own publicity, Shandong Ruyi is an 'award winning privately held textiles company employing approximately 40,000 people across China'. (*Cubbie Station excellence in cotton*, St George Information Bureau) It employed in Australia '40 farm employees on the Dirranbandi/St George properties and 35 ginnery employees during the ginning season.' Cubbie grows cotton, wheat, sorghum, barley, chickpeas, sunflowers, mungbeans and soybeans, which is exported.

Cubbie Station 80 per cent not Australian

Cubbie had secured a water licence for 460 gigalitres a year – the equivalent of 184,000 Olympic-sized swimming pools or the amount of water in Sydney Harbour. The station is often derided for its large water usage and the effect on the Australian river system. As late as December 2018, the $30 billion giant Shandong Ruyi refused to wind down its 80 per cent stake in Cubbie, despite being given more than six years to do so by Australia's elected government. (*Weekly Times*, 4 December 2018)

The Australian government announced in 2018 that the Foreign Investment Review Board approval was slashed from $252 million to

$15 million. This was perhaps to reassure the Australian people that this would reduce the ability of foreign investment in large properties and infrastructure. Sellers were now required to advertise holdings for thirty days before foreigners could apply. But more than ninety-two per cent of foreign-held agricultural land was held within Australian incorporated entities, which made the efficacy of such a regulatory process questionable and open to abuse.

Australia has a fragile ecosystem. The long-term effects of foreign investment and control of Australian agricultural land are unknown but if a nation loses its productive capacity, if a country cannot export what it once did for its own national profit, if it cannot feed its people as it once did, it is less a nation. The produce from multinational enterprises is exported with minimal financial gains to Australian gross domestic product. When the enterprises do not use Australian workers or enhance rural communities, the national loss is compounded. Farming families subjected to rushed foreclosures and fire sales of their properties wonder how complicit their banks and bureaucracy may have been in the transformation of and the control of Australia's scarce agricultural resources. There are too many questions and too few answers.

One farmer, Bill Mott, takes a sip of the amber fluid in front of him and pushes the remnants of the sandwich away as an alert on his mobile phone sounds. The sound of the phone is not entirely familiar in this outback with limited coverage and innumerable blackspots. The message is not from Thorburn. Instead it is a post from fellow bank victims. Thorburn was not even in Australia, but on a free holiday at a luxury Fijian resort where rooms cost between $6,000 and $4,5000 a night.

Bill Mott took one last mouthful of the shandy and walked out into the heat and blinding light of western Queensland.

Twelve

'It is only 41,000 kilometres around the world and I have done three times that chasing the National Australia Bank.'

The Western Downs visitor book displays a photo of a bumper wheat harvest; everything looks prosperous. A calendar below sets out the schedule for crops.

Winter cereals (wheat, barley, chickpeas) planting May and June; harvest October and November. Summer cereals (sorghum) September, October, November; harvest January, February and March

Oh, but if only it was so easy and predictable. It is still only spring 2018 and already creek after creek is in name only, their beds littered with dry withered growth. Once flourishing properties now lay waste, houses once full of laughter and the sound of children running down timbered floors are now silent and sad. The good times will come, so they continue to say. Queensland's *The Courier Mail* offers more pessimistic headlines: 'Bleak and dry outlook on farming' and 'Drought ravages GrainCorp'. 'October's rain too little, too late for some commodities...drought-ravaged east coast cropping landscape'. (*The Courier Mail*, 16 November 2018)

One farmer, however, wishes he was still fighting the elements on his land rather than fighting bureaucracy, bankers and bastards in cities. Bill Mott at least understood the harshness of farming and grazing on the Western Downs. He had repeatedly travelled to Melbourne, Sydney, Brisbane and Canberra to attend Royal Commission and parliamentary enquiries, hoping each month there would finally be the clarification and acknowledgement he had sought for years concerning the foreclosure on his family's farms. He had heard so very many words spoken, and few had enlightened him. He had heard promises made and none had come to fruition. 'It is only 41,000 kilometres around the world and I have done three times that chasing the National Australia Bank.' He had been promised more than once by NAB authorities that documents and resolution would come 'within a fortnight'. The latest had come from the NAB chief executive Andrew Thorburn himself, but that was weeks ago and Thorburn was now on a free luxury holiday in Fiji.

Financial failure was nothing new since deregulation in the 1980s. Millions of Australians had placed their trust in individuals and institutions which had promised care and ethical attention for their savings and livelihoods. Greed, bribery, corruption, poor compliance and disrespect had occurred too often and accountability and punishment too little.

The 1980s had revealed a scandal in Western Australia. The state government led by premier Brian Burke had engaged in business dealings with prominent businessmen such as Alan Bond, which led to an estimated outlay of $600 million of public money and the insolvency of large corporations. This had intermeshed with the Rothwells Bank scandal and the $1.8 billion collapse of Bond Corporation. An estimated $1.2 billion was said to have been siphoned from Bell Resources. The 1990s had been marked by the $3.1 billion State Bank of South Australia collapse. In 2003, $5.3 billion evaporated with the collapse of Australia's largest insurer, HIH. There had been federal government-commissioned reviews into the financial and banking sectors in 1991, 1996, 2010 and 2012. (*The Canberra Times*, 6 February 2019) What had been achieved, what had been resolved?

During this Royal Commission, NAB top executives had been openly criticised for their lack of empathy and acknowledgement of responsibility. They blamed the bank's culture on a few wayward staff. On 13 December 2018, the headlines read, 'Corruption Probe: Police raid home of ex-NAB officer'. (*The Canberra Times*, 13 December 2018) Fraud squad police had raided the home of Rosemary Rogers.

Victorian police assisted NSW detectives in executing search warrants at the Rogers home and seized computers and documents. This was part of a wider investigation involving suspicious payments by events and human resources firm Human Group, and its owner and director Helen Rosamond, to win contracts from NAB. Information had been uncovered that the luxury Fiji holiday and a Thermomix kitchen appliance, valued at $2,000, for Andrew Thorburn were paid for by the Human Group, though Thorburn was himself not under investigation.

On 5 March 2019, Rogers, the former chief of staff to NAB chief executives Cameron Clyde and then Andrew Thorburn, was charged with fifty-six counts of being an agent corruptly receiving a benefit, and dishonestly obtaining financial advantage by deception. The rich and famous lifestyle led by Rogers and her plumber husband, Anthony, had included a one-month family holiday to Europe worth almost $188,000, with an estimated $70,000 spent on accommodation alone; and a $485,000 New York, Washington, Miami and Hawaii holiday for eight people with flights costing around $150,000. There was a boat valued at $46,000, a $250,000 holiday to Fiji and multiple trips on private jets, helicopters, and luxury cruises. There were the bank cheques worth hundreds of thousands of dollars; prepaid credit cards, one of which was valued at $37,000; a $6.2 million property portfolio; three bank accounts; and a $1 million NAB bank cheque.

All were alleged bribes paid to Rogers between 2013 and 2017 for her role in the systematic NAB bank fraud worth $40 million – a fraud achieved through kickbacks and bloated invoices submitted to NAB by the Human Group. At the time of her arrest, Rogers was on holiday moving into a new trophy home valued at $3.8 million in Williamstown. It was believed Thorburn had summoned Rogers to his office on 11 December 2018 to query a transaction. She explained that the Human Group's Rosamond had lent her a bridging loan for the new house. At the end of the meeting, Rogers, who had worked within the NAB for two decades, resigned.

The lavish lifestyle should have initiated earlier questions and investigation within the NAB. It indicated not only very poor business principles within NAB but suggested a 'greed and corruption is acceptable if not good' culture, within an institution in which Australians trusted their lives. NAB released the statement that the bank was a 'victim of the alleged fraud, which, if proven, represented "a most serious breach of trust by a former employee"'. (*The Canberra Times*, 6 March 2019) This was a classic response, to blame an individual for a deep institutional quagmire. NAB executives had received exorbitant travel and luxurious accommodation freely. One luxurious Blue Mountains resort was the favoured location for executives to enjoy while discussing policy, including one ironically on 'cost-cutting'. (*The Weekend Australian*, 9–10 March 2019) The Human Group website proclaims that it can excel at

> Sharing company values & complimenting your ethos. Handle all matters with upmost integrity. An additional arm of your organisation with no hidden agendas. (http://www.humangroup.com.au/phone/index.html)

It was now revealed that NAB auditors had not examined the chief executive's office 'in any serious way for years'. Now, on examination, 'what they found was troubling'. (*The Weekend Australian*, 9–10 March 2019) The misappropriation of funds had been occurring for years. NAB insiders had expressed concerns that 'a culture of largesse and extravagant spending had become 'normalized inside Mr Thorburn's office'. (*The Sydney Morning Herald*, 2 February 2019) An estimated $113 million worth of corporate travel, bribery and kickbacks 'of NAB authorities' was believed to have taken place over a period of ten years. (*The Australian*, 6 February 2019) By 1 March, 2019, the NSW police announced it could not rule out further arrests.

> We will allege there are very senior positions involved…there are those in charge of carrying out business with integrity and it is very clear this hasn't occurred in this investigation. (*ABC News*, 1 March 2019)

The Mott family had paid eighteen per cent interest. They had repaid $6.5 million in interest and $550,000 in fees, which was equivalent to what they had originally borrowed, except the family still owed $12 million. Bill knew they had been charged around two and a half per cent interest more than they should have been; that contracts had been breached and they were defrauded of around $600,000 a year for six years. There had been about 'a dozen really bad things' done to the family by NAB. Initial enquiries to the bank had been met with the comment 'you would have gone broke anyway'. 'They broke us… When they realised what they had done, there was a fire sale to cover up the illegalities.'

Bill believed he had the information and most of the paperwork proving the manipulation of bank bills and how he had been deliberately set up to fail when he trusted his Roma bank manager, who assured him what a great product his loan was.

They stopped paying tax on the interest on our debt and they can then claim all interest as a write-off. I was sold the wrong product, and the bank bills and fees are equivalent to a term loan with redemption. Bank bills are supposed sold back to Treasury but if it is kept in house, they keep all the interest…isn't that fraud?

Even then, by devaluing the properties and selling in a hurry, the government was denied a greater value in stamp value.

Bill Mott had purchased a few NAB shares so that he could attend the NAB October 2018 general meeting and legally rise from the Melbourne shareholder audience to ask questions. When he rose to ask the board about the manipulation of bank bills and his rural debt, board members looked quizzically at each other and no answers came. Bill left six phone messages for David Gall, the NAB chief customer officer, the following day, and hand-delivered documents, but the silence was deafening.

NAB chiefs Ken Henry and Andrew Thorburn looked stunned in the photograph taken during the October general meeting. The shareholders had not just spoken, they had shouted their fury, venting their displeasure at the salary of Thorburn, asking what he could possibly do with $4 million per year and how, given the current banking turmoil, did he or anyone, deserve it. The words 'obscene and appalled' were heard from the enraged audience, who argued that instead of rewarding executives, they should be punished. Shareholders, eighty-eight per cent of them, voted against executive pay packets. Shareholders had watched as the NAB share price had dropped from nearly $30 to just above $23. They had been unimpressed by what had been revealed during the Banking Royal Commission. But in late 2018, shareholders were still unaware of just how bad it really and, by demanding greater profits, how complicit they too were.

By November 2018, more than a hundred bank, insurance and superannuation employees, including chief executive officers (CEOs) had fled their positions and the CEOs of the major four banks had yet to face their final government committee interrogation. It was clear that the commission had resulted in uncomfortable adjustments within the sector. During the past bruising year, NAB had spent $456 million on risk and compliance and a new committee was being set up to focus on customer outcomes. There had been candid answers concerning lack of concern for customers, and faint apologies and promises for a better culture prior to the Christmas 2018 break. Unfortunately, a whistleblower revealed that an email sent just before Christmas to staff from an executive requested they 'Knuckle down' to 'fill out funnel…our pipeline is low'. (Courtesy of Dan Ziffer, ABC) Staff were told they needed to sell mortgages and loans, to 'Crank it up', and that 'for the banker that can land "5 [applications] first this week"', there would be reward points redeemable for appliances and movie tickets. (Courtesy of Dan Ziffer, ABC)

The Hayne final report tabled on 1 February 2019 and its 215 documents detailed a decade of wrongdoing. Thousands of pages and detail, a slew of potentially criminal breaches by banks and their staff. Hayne condemned behaviour that had emanated from the executive suite to front line personnel – staff were measured and rewarded for sales and profits. The NAB came in for some of the worst criticism.

NAB stands apart from the three major banks. Having heard from both the CEO Mr. Thorburn, and the Chair Dr Henry, I am not as confident as I would wish to be that the lessons of the past have been learned. More particularly, I was not persuaded that NAB is willing to accept the necessary responsibility for deciding, for itself, what is the right thing to do, and then having its staff act accordingly. (Commonwealth of Australia, Royal Commission into Misconduct in the Banking, Superannuation and Financial Services Industry, Vols 1–3, 1 February 2019)

The commission made a mockery of the NAB mantra, EPIC – Empathy, Perform, Imagine, Connect. Victims believed the mantra EPIC came before the word FAIL. They were left to rue their belief in mutual trust with the realisation that they had been purely fodder for profit.

A generation raised to trust the banks had been exploited.

The recriminations were music to the ears of men like Bill Mott, but what did it all mean in the practical and here and now? The commission had laid bare bank misconduct, but its terms of reference had meant that it focused heavily on structural issues. There had been little ability to hear from but a token of men and women whose lives had been destroyed. For them, who had held such high hopes, it was another setback. One commented that the commission had

Acknowledged everyone except the victims…that was a kick in the guts and set the tone of what the report was going to be. (*The Sydney Morning Herald*, 14 February 2019

There was no scope for the commission to provide meaningful redress for thousands of Australians.

Clearly, the Australian Financial Complaints Authority had failed to protect those it was supposed to be protecting and the legal system had allowed banks to pressure customers into accepting unfair conditions. Commission Hayne's final report recommended that financial watchdogs consider criminal charges against organisations linked to the fees for no service scandal. The scandal, it was predicted, could cost wealth managers and banks around $850 million in compensation. While the treatment of fees for no services clients could not be diminished, the bank abuse within the rural sector was far greater. Any compensation could likely run into billions of dollars, not millions. Hayne rightly urged banks to ensure agricultural land valuations in future were conducted by individuals independent of lenders. Bank employees administering loans needed to be experienced agri-bankers, and the appointment of receivers or external administrators should be a last resort. Banks should be barred from charging dishonour fees on basic accounts. The banking code of conduct needed to be amended so that people in remote areas or those with poor English skills could access and conduct banking. Hayne believed banks should not charge default interest on farm loans in an area declared to be in drought or subject to other natural disasters. The commissioner believed that if banks adhered to these recommendations, it could result in a better outcome for farming families. The findings did nothing to resolve or appease farmers whose assets and homes had been discarded through bank and receiver fire sales.

The commission made seventy-six recommendations and the government committed immediately to just two: the establishment of a national farm debt mediation program and a last-resort compensation scheme to those whose financial providers had gone into liquidation. For farmers, a national farm debt mediation program meant more rhetoric, more paperwork, more protracted negotiations in which they had little power, when they were already broke and so very weary. The compensation scheme meant nothing because rural lenders were still operating.

The Australian government had fought against the conduct of a Royal Commission until no other political option was left and had then imposed strict terms of reference and a tight time frame. Prime Minister Morrison as treasurer had voted against an inquiry into the banks twenty-six times and was, according to reports, the last in cabinet to agree to the commission. (*West Australian*, 5 February 2019) 'In opposing the royal commission so stridently and for so long the Government has entrenched itself on the wrong side of history', wrote one reporter. (*West Australian*, 5 February 2019) This resulted in accusations of a government which favoured the 'top end of town' at the expense of the general public. Current treasurer Josh Frydenberg refused to concede the government had been wrong in its opposition and declined to apologise.

The refusal of Commissioner Hayne to shake hands with Treasurer Frydenberg and smile for the media after delivering the three-volume report may well have reflected how he felt about the restrictive conditions the commission had faced. This was demonstrably shown in the surge in bank shares value the following day. ANZ shares rose 6.5 per cent, NAB 3.9 per cent and CBA 4.7 per cent. One banking analyst referred to the report as 'tough talk, soft recommendations', another, 'clear win for banks'. (*ABC News*, 5 February 2019)

The banks had feared Hayne would recommend tighter loan assessment standards – he didn't, or couldn't, do so. Analysts with the investment bank UBS described the Hayne final report as a clear win for the banks. Without strong recommendations, could lasting cultural change be assured, particularly as those who sat on management boards commonly rotated? The inevitable questions were 'could they be trusted to do so?' and could self-restraint be nurtured successfully in a culture of greed, in an industry measured by sales and profit?

Hayne recommended that APRA and ASIC be more proactive in

launching legal action against large corporations, rather than simply issuing infringement notices if breaches were witnessed. He believed the Federal Court needed expanded powers to cover corporate criminal misconduct. The Australian government countered with 'much of the change needs to come from the regulators themselves'. (*ABC News*, 4 February 2019) The dozens of references within the report to misdeeds could beyond doubt lead to the civil and criminal prosecution of individuals and entities but many Australians believed that, given the influence and power held by those within the top echelon of the financial sector, any such prosecution would likely be contested vigorously until the afflicted had no more time, money and energy left.

Thirteen

'I've just about gone to the ends of the earth with this. I've done everything available to somebody in my position.'

It had been an information overload but those responsible for unethical and uncaring behaviour continued to deny responsibility as victims continued to struggle.

The Parliamentary Select Committee on Lending to Primary Production Customers had released its report in December 2017. It was sympathetic to farmers, damning of the financial sector and supportive of further investigation. The reaction from the Australian Restructuring Insolvency & Turnaround Association (ARITA) was immediate and condemning of the committee and report.

> The Report displays an abject misunderstanding of insolvency law and of the work that receivers are required, by law, to carry out, and blind acceptance of the views of insolvent farmers. (www.arita.com.au/ARITA/News/ARITA_News/Senate-Inquiry-gets-it-wrong-on-rural- receiverships.aspx)

ARITA chief executive officer John Winter believed the committee sided with 'completely disproved claims of some insolvent farmers'. He defended receiver fees as not being 'excessive' and declared only 'a handful' of receiverships each year were agriculture-related and 'fewer still relate to family-owned farms'. Many of the recommendations made in the rebuke were reasonable, though the emphasis was on the poor lending and borrowing practices rather than any fault within the insolvency sector. It was also an unnecessarily harsh generalisation to state,

> In almost all of these cases, the receivers were found to have brought in farm management experts to attempt to turn around the poor farming practices and poor animal welfare management that had contributed to the need for their appointment. Unfortu-

nately, the insolvent farmers persist with blame-shifting for their own failings. (www.arita.com.au/ARITA/News/ARITA_News/Senate-Inquiry-gets-it-wrong-on-rural-receiverships.aspx)

The banks had long maintained their silence. Through the efforts of minority party and independent politicians, the call for a major investigation into financial institution misconduct grew louder, but the Australian major parties had resisted. It was left to a bloke named Jeff Morris to disclose what should have been disclosed by his government and its agencies. Jeff Morris had started work with the Commonwealth Bank in 2008 as a financial planner. He believed that banks were trustworthy institutions intent on treating their customers with care. He quickly became aware of corruption going 'all the way to the top'. He was appalled.

Corruption was absolutely rife. There were crooked plans stitching up literally widows and orphans, and there was a crooked management team covering it up. (https://www.jeffmorris.com.au)

Morris believed everything deteriorated from the late 1980s due to a fallout from foreign currency loans which were marketed to unsophisticated borrowers such as farmers for whom such loans were totally unsuitable. (*ABC News*, 30 November 2017) Banks then covered up and blamed the customer. Morris and two other concerned staff decided to report the misconduct to the financial regulator ASIC. ASIC was 'useless' and 'did nothing'. (*ABC News*, 30 November 2017) He felt he had no alternative but to go to the media and then appear before a parliamentary hearing. Morris lost his job, was vilified, and his safety and that of his family was threatened.

The day after the Hayne Royal Commission February 2018 report was released, bank stocks soared giving the Australian share market its biggest one-day rise in more than two years, with $19 billion dollars poured back into the banking sector. Although Hayne had been adamant that the gap between the NAB-promoted public image and the way it undertook business was extensive, NAB shares rose. Bank victims could do nothing but lament what could, should, have been.

The media photo appearing on 6 February, 2019, was of a smiling, waving NAB chief executive, Andrew Thorburn. The headlines read 'Banks enjoy Hayne bounce' and 'NAB bosses dig in: we'll fix the culture'. (*The Australian*, 6 February 2019) The day after pledging 'we'll fix the culture', Thorburn and chairman Ken Henry resigned with 'deeply sorry' comments. Henry would remain until a new chairman was appointed. Thorburn had been fourteen years with NAB, having joined the bank in 2005 as the head of retail banking. He was promoted to the top position in 2014. From 2005, the NAB had conducted some of its most questionable customer dealings. Though his salary had been cut, Thorburn left with a settlement of $4.3 million.

The Commonwealth Bank admitted on 9 March 2019 that it had paid $1.4 billion in remediation during the previous five years. The NAB was forced to concede it had underestimated customer compensation. Originally it believed 'just a few million' was required for compensation. By April, the bank agreed to pay out to customers an additional $525 million after tax, blowing out the customer-related remediation program cost to $1.102 billion. (*The Financial Review*, 18 April 2019) It was suggested that the total compensation paid by all four banks could rise to $10 billion.

Bill Mott continued to fight for justice. Yes, the money remained an issue but not *the* issue. 'It is not right and needs to be fixed.' There were so many interwoven beliefs and grievances. It had taken over his life and he knew that, but he also held firm to the conviction that he was right, and his family had been defrauded. So many others were exhausted, emotionally and financially, and had walked away – their lives and futures shadows of what they once were. Bill tenaciously hung on because 'the bastards shouldn't be allowed to win', but he admitted redress from the NAB was an epic journey. 'Every time you get something positive, some momentum happening, they derail you.' The unwillingness on the part of NAB to settle was beyond belief. The bank had spent so much money contesting resolution with the Mott family that it was nonsensical.

Bill repeatedly drove between Meandarra and Melbourne, Sydney,

Brisbane and Canberra, staying in cheap hotels. The receptionists at the Melbourne NAB headquarters greeted him by name but he never made it past the foyer. The NAB ensured his self-funded travel continued when he was informed former Victorian premier, Jeff Kennett, had been employed by the bank to arbitrate individual cases and this meant more interstate trips to face Kennett and two bank employees.

Emotions were confused by the news. Kennett had not been trained in arbitration. In 2013, he had referred Coles to the competition regulator over false 'baked fresh' bread claims. 'Two years later, he was adjudicating between Coles and its smallest suppliers.' (*The Australian*, 1 February 2019) Kennett was being paid by the NAB and the process had been an initiative of Andrew Thorburn, so there was doubt. Kennett had nonetheless founded Beyond Blue in 2000, an organisation to raise awareness of depression and anxiety and to prevent suicide, and had led the organisation for seventeen years. This, it was believed, should make him sympathetic to the mental and emotional trauma endured by NAB victims.

In his first meeting with the arbitration panel, Bill conceded Kennett seemed 'a decent bloke'. Kennett examined the documents and listened and then promised a result 'within a fortnight'. Bill was asked to attend another arbitration meeting, in another state – more self-funded travel. Kennett brought down his judgement – 'it was a ridiculous offer'. Bill was more than angry – yet again, the NAB had stymied him. By mid-January, any optimism in Kennett's arbitration was completely eroded – the compensation offers were nothing short of insulting, and twenty-one of the twenty-five complainants had refused to be further involved. The collective opinion was that Kennett 'wasn't truly independent, was harsh with people and was out of his depth in complex banking cases'. (*The Australian*, 1 February 2019) They were left to ponder how much money yet another non-productive, timewasting process had cost NAB. It had been 'a bitter disappointment' (www.bankreformnow.com.au) and it was back to the future.

In a near unanimous decision the victims chose to revert back to dealing directly with NAB executives who had a much better understanding regarding the deep-inner workings in their bank. (www.bankreformnow.com.au)

The promises made by Andrew Thorburn had been untrue. He had assured Bill Mott the bank documents dealing with the Mott farms would be forthcoming 'within a fortnight', that settlement would be forthcoming 'within a fortnight' – neither happened. Thorburn had jumped ship with a golden handshake, while the NAB, for all the mumbled apologies, seemed slow to change, and acted relentlessly negatively in dealings with these ordinary Australians. Dr Peter Branson, founder of the group Bank Reform Now, believed,

> Unfortunately, the bank seems to have attempted to revert back to the now defunct – Deny, Delay, Deceive routine. These cases could have been sorted out in days or weeks, but salt has been rubbed into the wounds with a process that gutted victims over the last three months. (www.bankreformnow.com.au)

The NAB promise to improve its public image and be more caring could have been enhanced by settlement. That opportunity was lost, and the top officials faced yet another Australian Parliamentary Committee interrogation in Canberra.

It was now March 2019 and Bill Mott and other Bank Warriors were back in Australia's capital for the next House of Representatives Economic Committee hearing – the same politicians, facing a new NAB chair, Philip

Chronican. Bank Warriors were sceptical that the new Chair could bring an improved attitude to their cases. Chronican, a current NAB director, had more than thirty-five years of experience in banking and finance in Australia and New Zealand. Before his positions with NAB, his long career included group executive Westpac Institutional Bank and chief financial officer; and then responsibility for ANZ's retail and commercial businesses. He clearly was well qualified in Australian banking, but Australian banking had clearly been obsessed with earnings, revenue and growth, while disregarding the methods with which this was achieved.

Before the hearing on 27 March 2019, NAB public relations was busy. Media headlines included 'NAB to revamp pay in bid to improve culture'. (*The Sydney Morning Herald*, 26 March 2019) Mike Baird, former NSW premier, and now head of NAB's consumer banking arm, admitted, 'we have to respond' to the poor outcomes witnessed in the banking culture. (*The Sydney Morning Herald*, 26 March 2019) Chronican was quoted as saying,

> Much needs to change in a meaningful way at our bank. We have unfortunately been found wanting in too many areas when it comes to our customers, and I am determined to ensure that change happens to ensure we meet and exceed their expectations. (https://news.nab.com.au)

It had been a long road for Bill Mott as, yet again, he took his seat in the main committee hearing room of his nation's Parliament House. It was increasingly difficult for him to believe he was any closer to restitution than he had been years ago. The anger was not healthy, but it was anger that fuelled him, and the belief that 'It had less to do with profit than just covering their arses.'

The hearing progressed slowly with committee questions, NAB measured answers, and an audience struggling to remain silent. 'No, there was no default interest to agricultural customers', was met with muttering. 'Why had there been no accountability?' 'ASIC is still considering prosecution, what about accountability?' An answer: 'Bank staff were encouraged to practice self-assessment.' 'Had they not been encouraged to practise self-assessment previously?' Not answered. From the committee there came the suggestion that a more fearful cohort was needed to enforce bank good behaviour. Fees for no service was raised. NAB executives admitted it was a badly implemented policy but that the NAB had not knowingly or intentionally set out to deceive customers. An admission was made that over the previous five years around $800 million remediation payments had been made by the NAB.

Chronican mentioned that the way advisors left and new ones were appointed may have resulted in a lack of consultation. From the committee, 'Why was a whistleblower necessary rather than good internal risk assessment?' The answer: 'Whistleblowers were part of NAB risk management.' A whistleblower had broken the silence in December 2017, resulting in an NAB investigation in March 2018. This was followed by forensic accountancy, with the results being disclosed to the CEO in September 2018. There was a discussion between the panel and NAB executives concerning the difference between 'illegal' and 'charging something that is illegal'. 'Has anyone been held accountable?' There was an admission that NAB staff had been terminated. No one went to jail. 'Why was not an executive held responsible?' No answer. The hearing had gone on for more than an hour and the largest disagreement had been about the specific definition of words and phrases. Only one television cameraman remained.

The dancing with words continued, the language unnecessarily prolix. 'How many outstanding complaints remained?' The executives conferred and believed out of the 1,064 complaints received since November 2018, 559 remained outstanding. The Banking Reform Now members sat in the front row to make their presence visible to the media, their black and red T-shirts impossible to miss.

Unfortunately, by the two-hour mark of the wordy hearing, the media ws bored, and few remained.

A parliamentarian raised concern about the decrease in Australian house prices and how this could affect house value, equity and mortgages. There was no enlightening comment from the bankers. Another member of the committee commented on the way interest doubled with impairment and possibly doubled again once a loan was in default. The manipulation of bank bills was raised but there was no admission from the NAB executives, even though ASIC had fined NAB $50 million. Chronican assured the committee that if he came across such manipulation, he would advise committee members. The bankers acknowledged that after the 2008 global financial crisis the banking industry became very complicated. They glanced at each other and confessed that some forty-four staff had been recognised as being in breach of bank protocol; twenty-one were 'terminated' and twenty-three received counselling. Furthermore, of the $800 million remedial payments made, most was paid out of shareholder profits and 'some out of staff wages'.

The hearing was now in its third hour and it was difficult for those present not to be annoyed by the profuse rhetoric and clear stalling. The audience became restless when it was suggested that full interest on agricultural loans had not been expected in times of drought and few farmers had got into trouble – 'the NAB was working with them to resolve and in a cordial manner'. Bill Mott squirmed in his seat when executives attempted to paint the NAB as a benevolent institution. It had been reported that major banks had taken an average of 1,726 days – more than four and a half years – to identify significant breaches, and on average a further 150 days to notify ASIC that an investigation had commenced. The Corporation Act required banks to report breaches to the ASIC within ten days. (*ABC News*, 25 September 2018)

The NAB chair agreed that culture came from the top, but there was no magic wand. Employees were being encouraged to speak up. 'Will it settle down to what it was?' 'No,' replied the NAB chair. Someone in the audience murmured,: 'How will there be a different culture with the same bankers in control?' Chronican was asked how long it took for disputes to be resolved. Most, apparently, occurred within five days. Bill shook his head. Chronican was asked how long it took the NAB to hand over requested bank documents to plaintiffs? 'A week perhaps, two to three weeks at the most.' Bill almost choked. 'Why has it been more than six months and I'm still waiting?' he whispered but wanted to shout.

A subdued and casually dressed man took photographs during the proceedings. He was another bank victim. The Bank Warriors helped where they could. This bloke had lost everything and slept in his car until he found a beaten-up caravan at the tip, hauled it out and fixed it up as best he could. Bill and others slipped him money for fuel.

Another who slept in his car so he could attend hearings was Craig Caulfield. In 2007, Craig Caulfield and his wife obtained a loan through a Commonwealth Bank broker to purchase a 110-acre Queensland cane farm. They fell behind in their repayments and in 2010 the Caulfields approached local branch CBA staff and asked for advice. Staff contacted CBA head office. The couple were told to wait for a letter. The letter arrived and they were shocked to see it was from a CBA solicitor. Full loan repayment plus arrears was requested immediately. 'We were unable to make such a payment and we were obviously very worried.' The Caulfields made an offer to pay the arrears and fees in full and then repay the loan gradually. This was immediately refused. They were informed they were not legally entitled to mediation. Reviewing loan documents, Craig discovered massive errors which he believed included data fraud and even forgery to protect the bank from charges of predatory lending. He presented the bank with twenty-seven incorrect and fraudulent entries. The CBA refused to negotiate or mediate. Craig Caulfield's health deteriorated, along with the family's financial situation. Their living situation became dire and Craig attempted suicide. His wife was diagnosed with a massive growth in her abdomen.

Craig Caulfield believes his fellow Australians 'are still unaware of the depth of destruction'. He had felt alone until he met other Bank Warriors; their mateship and support meant a great deal. He did not now intend to give up in any way.

Neither did Bill Mott. He had lost his family, community, home, livelihood, farm and future. He was up against an enormous and powerful entity. Litigation was out of the question. Not only was he broke but the bank had access to limitless resources. So too could it use the legal system to bludgeon victims into submission. The bank had most of the documents and could ruthlessly exploit this by denying access – documents could be 'lost' or 'difficult to locate'. Bill Mott was still waiting for his. The enquiries, hearings, Royal Commission and personal effort had vindicated Bill's belief that he had been defrauded but it had taken years of his life. 'I've just about gone to the ends of the earth with this. I've done everything available to somebody in my position.'

Fourteen

'I think they hope I might die in the meantime.'

Tales of unadulterated greed, bribery, corruption, poor compliance and blatant disrespect had left Australians bewildered. The institutions which had demanded respect for generations had been found untrustworthy. It was increasingly difficult to comprehend the void between the present and the most common historical interpretation of Australian culture.

In April 2019, crowds larger than the year before gathered for Anzac Day services and stood for hours to applaud military veterans. The same month, thousands thronged through the gates of Sydney's Royal Easter Show, not just to enjoy sideshow amusements but to appreciate and support Australian farmers and produce. *The Royal Agricultural Society Times* magazine promoted the past, the front cover adorned with women in period costume. It was full of optimism, wonderful smiles, bountiful harvest and pride – and not one non-Anglo-Australian face. The emphasis was that the '2019 Sydney Royal Easter Show will take you back to the future'. (*Royal Agricultural Society Times*, March 2018, Vol. 17-1) Yet the previous months had revealed a flood of information counterposed to this carefree, prosperous and iconic past.

Royal Agricultural Society Times, March 2018

Bill Mott was inexorably tied to the past, as he was to this confusing present and unfathomable future. His grandfather William Gordon Mott had spent his youth fighting on Europe's Western Front in the war that was supposed to end all wars. Wounded in body and mind, he returned to rural Australia. His was a generation scarred but shaped by war and the ensuing Great Depression. They were hardworking and frugal with fundamental faith in the church, community, government and institutions. All they asked for was 'a fair go'. These beliefs they handed down to their children and grandchildren. William Mott fought harsh living conditions in Queensland's unforgiving west. His grandson, Bill, asked for nothing more. He too was a man of the land who believed in hard work, family, community, government and institutions. By the beginning of the twenty-first century, despite the hard work, the beliefs were in tatters.

For decades, bank customers had been vigorously persuaded to accept inappropriate products and loans larger than they could afford by bank staff incentivised to grow business regardless of the human toll. Months after the Royal Commission, bank executives remained coy about the more than twenty-four cases referred to the regulator for potential prosecution. The 'failures of process, failures of judgment, failures of process, failures of leadership and in some instances greed' (*ABC News*, 22 October 2018) had been proven in late October 2018, but Bill Mott was still chasing the NAB for settlement.

Then there was a whiff of optimism when in April he and Lynton Freeman were invited again to Melbourne NAB headquarters, this time to meet with Ross McNaughton, NAB general manager, strategic business services, and Craig Swinburne, general manager, business customer services. Finally, this was it – the long journey to resolution was here. They were confident the forensic analysis of the bank bills and loan manipulation was the proof needed. The meeting was convivial, and the officials listened. Bill was asked if he would hand over copies of the forensic and other accounting analysis, and the bank would get back to him 'within a fortnight'.

There was no communication within a fortnight, so Bill made the phone call. The conversation was brief. NAB had engaged Price Waterhouse Accounting to examine the Mott family documents he had provided, and Bill would be notified when this worldwide accounting colossus came up 'with a settlement within a week'. The NAB likely was to be charged in the vicinity of $1,500 an hour by any number of professional accountants to settle with a farming family they had foreclosed on. 'That is what they go to deploy, to beat a farmer.' It was another kick in the guts for a bloke who had only ever wanted to work the land. It wasn't a week or two – another month dragged into another.

The same month that Australians commemorated Anzac Day and enjoyed the Sydney Royal Easter Show, there was another rural ecological, bureaucratic conundrum. The Murray-Darling River system had long reflected the fragility of the Australian landscape and more recently the harshness of climate change. Water levels were at an all-time low and Australians were shocked by news footage of millions of dying fish in the Menindee Lakes region. The following day, mismanagement and possible corruption of the Murray-Darling Basin Plan was disclosed.

The conversation between agricultural communities and environmentalists concerning how much water should be syphoned from the river system had always been fraught, in this the driest continent. In 2008, the Australian Labor government implemented a scheme to ensure the waters flowed. The Murray-Darling Basin Plan limited the volume of water farmers could extract from the river, while a government buyback scheme was instigated. Under open tender, agriculturalists could register the price they were prepared to sell back water rights and the government department responsible decided on the best offers.

In April 2019, it was revealed that since 2015 the government had purchased surface water by direct negotiation with holders. The high prices paid in August 2017 to vendor Eastern Australia Agricultural raised questions. (*ABC News*, 23 April 2019) It was then found that the Australian minister for energy, Angus Taylor, was director and secretary of Eastern Agriculture between mid-2008 and late-2009. He was also listed as 'co-founder and director, Eastern Australia Irrigation from 2007'. (*ABC News*, 23 April 2019) Eastern Australian Irrigation was registered in the Cayman Islands. Centre Alliance Senator Rex Patrick, through the Senate Estimates processes, had discovered the limited tender, and offshore beneficiary. He believed, 'It's totally unacceptable that the Commonwealth uses taxpayer's money to transact with companies that are based in tax havens.' (*ABC News*, 23 April 2019) The minister for water in 2017 was Barnaby Joyce. In 2019, he said he had not been aware of who benefited from the buyback in 2017, nor had he been involved in the purchase price. Belatedly, the government admitted that transparency had not been 'systematic'.

On 4 June, Bill Mott and Lynton Freeman were finally summoned to NAB Melbourne headquarters again – at their own expense. They were subjected to an hour-long presentation of the findings of Price Waterhouse. Bill and Lynton were astonished. It took them a lot less than an hour to disprove the findings. Bank authorities conceded the findings were flawed. Another meeting in Melbourne was planned 'in a fortnight'. 'It is frustrating' that the bank would simply 'not take responsibility'. He was promised that the next meeting would be over two or three days to properly examine all the documents with bank staff so a settlement could be reached. It was difficult to believe a settlement would be achieved in a fortnight but at least for the first time NAB agreed to pay his travel expenses. It was happy hour and Bill intended to have a couple of beers – at bank expense. 'I could get used to taking money from the bank.'

The same month, unusually, daily newspapers included a thirty-nine-page *AgJournal* magazine supplement. The headline on the front was 'We reveal the nation's biggest landholders'. The photos were predominantly in colour though those featured throughout were not. Everything looked wonderful in the rural sector, and very white Australian. It was impossible not to question the timing and validity of such a free and widespread supplement. It broadcast that Australia was owned by white Australians for the betterment of Australians. Only one of the twenty largest agricultural empires mentioned an overseas partnership.

By the third week in June 2019, there was still no good news for Bill Mott and the agricultural and business news continued to be disturbing. Graincorp, Australia's largest grain handler, announced a $59 million half-year loss due to drought in Australia's east reducing production by 45 per cent. Australia is one of the largest grain-producing nations in the world but on 18 June a shipload of Canadian wheat was being unloaded at a NSW port. It was the first time Australia had imported grain in more than a decade. It was also announced that the national cattle herd had shrunk to its lowest numbers in twenty-five years. (*ABC News*, 18 June 2019)

Other news included the revelation that the ANZ New Zealand Bank chief executive had left the bank 'after an internal review raised concern among board members over his personal expenses'. David Hisco had been with ANZ NZ for more than twenty years and though he rejected some of the board's concerns, he 'accepted accountability and forfeited all unvested equity rewards'. (*The Canberra Times*, 18 June 2019)

For the agricultural sector, there had been too little understanding and poor support from government agencies. The emphasis was on revenue, trade, diplomatic expediency and political preservation. Rural communities already under pressure from industrialisation and globalisation began to diminish and disappear. Further quick-fix revenue-raising combined with diplomatic and political manipulation led to rapid foreign ownership of Australian natural resources and declining food and water security. Complicit were financial institutions and a legal system geared towards corporate profit rather than the welfare of individual Australians.

It is September 2019. There have been more meetings between Bill Mott and National Australia Bank executives. It is now five years this farmer has battled for justice. It has been five years of frustrating prevarication from the financial institution which took his livelihood, his home and his children's inheritance, and destroyed his marriage. At the last meeting, he was told mediation and discussion with lawyers was required – Bill Mott has heard it all before, too many times. 'I think they hope I might die in the meantime.'

Postscript

It is March 2020, a new decade – same battle!

Bill Mott is, in his own words, as tough 'a bastard' as you could meet. He grew up tough, required to slip out of his education to take over the running of the family Queensland property. There were no favours given, nor expected. That was, and continues to be, farming in Australia. The harshness of seasons was anticipated but along the way the failures in bureaucracy and bankers weren't. He had just had to get on with it. There was not choice really – you walked off the land or you weathered it, in every sense of the word. Bill always wanted to be a farmer, there was nothing else, so he stayed and got on with it. Floods, drought, fire he took in his stride, but the failure of the wool industry and other bureaucratic bungles proved more problematic – there was no control and little sympathy.

He believed the National Australia Bank manager was his friend, had his best interests in mind. That was a mistake he never saw coming. Fraud was involved, the 'this is business' and 'protect the bank at all costs' approach triumphed over humanity. Bureaucracy failed again and the bastards moved in. The Mott family lost everything almost overnight. It was around $20 million or so but losing your home, livelihood, self-respect and future – no price could be put on that. His children needed to move away, and his marriage ended. He could have walked away but the 'bastard' emerged and all the energy and long hours that farming took turned into a fight against the bankers and bastards.

Every denial, obstruction, procrastination, financial barrier needed to be overcome. The meetings, hearings, mediations were too many to count. The thousands and thousands of kilometres driven broke vehicles. By Christmas 2019, Bill Mott believed he too was very close to breaking. The five-year battle had taken a huge toll not only financially, but physically, mentally and emotionally. 'It was tough by Christmas 2019.' That is about as sad a confession as you could ever hear come from this farmer's lips.

He was attempting to take the bank to yet another mediation. He had been pretty sure he had all the Ts crossed – but he had before. This time he was stretched to the limit because the queen's counsel was going to cost him $120,000 for three days in court. He had engaged the best in Queensland, but the bank's law team were more in number and very experienced in protecting the bank and casting blame on the victim.

Bill Mott won. In March 2020, another decade, liability was admitted. After almost seven years, he had been vindicated. The financial settlement was 'nice' but the battle had always been about more than that, because the 'Bureaucracy, Bankers and Bastards' should never be allowed to win.

www.ingramcontent.com/pod-product-compliance
Lightning Source LLC
Chambersburg PA
CBHW070916080526
44589CB00013B/1323